COMPARATIVE BEHAVIOR OF BEES AND ONAGRACEAE

V. CAMISSONIA AND OENOTHERA BEES OF CISMONTANE CALIFORNIA AND BAJA CALIFORNIA

COMPARATIVE BEHAVIOR OF BEES AND ONAGRACEAE

V. CAMISSONIA AND OENOTHERA BEES OF CISMONTANE CALIFORNIA AND BAJA CALIFORNIA

BY

E. G. LINSLEY, J. W. MacSWAIN, P. H. RAVEN and R. W. THORP

UNIVERSITY OF CALIFORNIA PRESS
BERKELEY • LOS ANGELES • LONDON
1973

Volume 71

Approved for publication August 7, 1972
Issued April 1, 1973

University of California Press
Berkeley and Los Angeles
California

University of California Press, Ltd.
London, England

ISBN: 0-520-09474-3

Library of Congress Catalog Card No.: 72-619638

printed in the united states of america

COMPARATIVE BEHAVIOR OF BEES AND ONAGRACEAE

V. *CAMISSONIA* AND *OENOTHERA* BEES OF CISMONTANE CALIFORNIA AND BAJA CALIFORNIA

BY

E. G. LINSLEY[1], J. W. MacSWAIN[1,2], P. H. RAVEN[3], and R. W. THORP[4]

THIS IS THE FIFTH of a series of studies on the interrelationships of bees and evening primroses in a limited geographical region. The areas covered are more diverse ecologically than in the first three reports on the Colorado Desert, the Great Basin, and the Mojave Desert (Linsley, MacSwain, and Raven, 1963*a* 1963*b*, 1964). Emphasis has been given to the 18 species of andrenid bees which are oligolectic on these plants in these areas.

The associations of insects with flowering plants of the family Onagraceae are of considerable interest because of several prominent morphological attributes of this family. The pollen grains are relatively large, often more than 100 microns in diameter, and this alone makes them relatively difficult for bees to handle. In addition, they are bound together by slender viscin threads which make them adhere in masses. Plants of this family have an inferior ovary and a tubular hypanthium (calyx-tube) at the base of which nectar is secreted. This makes many members of the Onagraceae excellent moth and bird flowers, depending on the associated flower colors and odors, but poses a difficulty from the standpoint of bees, which must develop relatively elongate mouth parts if they are to exploit the nectar secreted into this tube.

In addition to the specializations characteristic of the family, which would in themselves insure a high degree of oligolecty on the part of bees gathering pollen from them, the genera belonging to the tribe Onagreae are even more specialized. As can be deduced from some of the common names associated with the members of this group—"evening primrose," "sun cups"—members of this tribe tend to have highly synchronized diurnal times of opening. Thus a genus such as *Oenothera* basically opens its flowers near sunset, whereas a genus such as *Camissonia* basically opens them near sunrise. In both cases, the flowers last for less than a day and are replaced the following day by a new crop. As a consequence of this short duration, the flowers which are pollinated must be pollinated within a very short time of their first opening and this also insures specialized pollination systems. This is true both because the insects which visit the plants must operate in a highly specific period of time, in which competition between potential flower feeders may become very intense, and because if they do not operate efficiently within this time, pollination will not be effected. In only three genera of the tribe Onagreae (*Gongylocarpus, Clarkia Calylophus*), as summarized in the preceding paper of this series (MacSwain, Raven, and Thorp, 1973) do the flowers last more

[1] Division of Entomology, Department of Entomological Sciences, University of California, Berkeley, California, 94720.

[2] Deceased 9 September 1970

[3] Missouri Botanical Garden and Washington University, St. Louis, Missouri, 63110.

[4] Department of Entomology, University of California, Davis, California, 95616.

than a day and even in these, the times of floral opening are highly synchronized.

For all of these reasons, we have found the members of the tribe Onagreae excellent objects for study in terms of pollination. We have found many oligolectic bees, with varying degrees of specialization, associated with them in different parts of the western United States. The present study summarizes the relationships between two genera of this tribe—*Camissonia* and *Oenothera*—in cismontane California and adjacent Baja California.

Acknowledgments

This is one of a series of related studies supported by grants from the National Science Foundation. The authors are indebted to P. H. Timberlake, who made or confirmed most of the identifications of bees; G. E. Bohart, who identified the *Dufourea;* and J. A. Chemsak, J. K. Drew, D. P. Gregory, P. D. Hurd, Jr., Juanita M. Linsley, A. E. Michelbacher, C. W. O'Brien, J. A. Powell, J. R. Powers, R. R. Snelling, G. I. Stage, and C. A. Tauber (née Toschi) who assisted in the collection of field samples. Insect samples taken during this study are preserved in the collections of the California Insect Survey, University of California, Berkeley. Plant specimens are in the herbaria of Rancho Santa Ana Botanic Garden, Claremont, California, and the University of California, Berkeley.

LOCALITIES WHERE STUDIES WERE CONDUCTED

Our studies were carried out in the following 47 localities. California. Alameda Co.: San Leandro Hills, Oakland. Contra Costa Co.: El Cerrito. Fresno Co.: 6.4 miles north of Coalinga. Kern Co.: 7 miles north of Bakersfield (on road to Woody); 18 miles east of Bakersfield; 9.6 miles north of Blackwells Corner. Lake Co.: 5 miles north, and 2 miles southwest of Lakeport. Los Angeles Co.: Claremont; Point Dume. Madera Co.: 3.9 and 7.6 miles south of Coarsegold; 3.7 miles west and Oakhurst. Merced Co.: 2 miles south, and 2.5 miles southwest of Livingston. Monterey Co.: 1.5 miles west of Bradley; 5 and 7 miles northeast of Carmel; 1.5 miles southeast of Carmel; Jolon; Santa Rita. Riverside Co.: 7 miles east of Temecula. San Benito Co.: 1 mile west of San Juan Bautista. San Diego Co.: 2 miles south and Carlsbad; 2 miles northeast of Lakeside. San Francisco Co.: Bakers Beach, San Francisco; Fleishhacker Zoo, San Francisco. San Luis Obispo Co.: 1.8 miles west of Bee Rock; Cambria; 2.6 miles south and Creston; Grover City; 4.5 and 5 miles northeast of Santa Margarita; 5 miles and 6.8 miles southwest of Shandon; 10 miles west of Simmler; San Mateo Co.: San Bruno Mountains; 1.7 miles southeast of Woodside. Santa Barbara Co.: Goleta; 1.6 miles west, and 2.6 miles east of Pine Canyon Guard Station. Santa Clara Co.: Jasper Ridge. Stanislaus Co.: Del Puerto Canyon; Turlock. Tulare Co.: 1 mile north, and 4.5 miles south of Tulare; Tulare Airpark (south of Tulare); 2.4 miles south of Tipton. Ventura Co.: 2 miles south of Point Mugu; south end of San Buenaventura Beach State Park; Hungry Valley, 5 miles south of Gorman. Baja California: 3 miles south of Ensenada; 7 miles north of Santo Tomás; 14 miles south of San Quintín. Several of these localities or categories of localities are worthy of special note and are briefly characterized in the following paragraphs.

The locality 18 miles east of Bakersfield is on the old highway over Tehachapi Pass. This is an area of fine wind-blown sand that has accumulated as a steep, flat-topped hill. The top of the hill is covered in favorable years by a dense stand of *Camissonia campestris* while the steep south-facing slope is heavily populated by *C. boothii decorticans.* There was moderate to dense bloom of *Camissonia* in 1958 and again in 1960, 1961, and 1962. There was almost no bloom in 1959 and 1963. This area is now largely under cultivation and may soon become unsuitable for colonies of *C. campestris.* This was our most intensively studied site since nine oligolectic *Andrena* species occurred on the two camissonias.

The seven localities along Highway 99 through the Central Valley of California are the only areas where small populations of *C. campestris* could be found. Each of these sites may be characterized as a small plot of unleveled fine sand surrounded by level agricultural land (pl. 4). Most of these have now been destroyed for various practical uses. However, around the southern margin of the Central Valley there are occasional small areas of fine sand which still maintain populations of this plant species and its bee pollinators, although their future is also uncertain.

The coastal localities are all areas of compacted sand, above the beach line, which support *Camissonia cheiranthifolia cheiranthifolia* along the north coast and *C. c. suffruticosa* along the south coast (pl. 3). Almost every area which had populations of this species were found to have oligolectic bee species. In these coastal situations this perennial plant produces numerous flowers every year over a period of many months and the oligolectic bees exhibit unusually long periods of activity.

The locality 14 miles south of San Quintín in Baja California consists of dunes of fine white sand. Here there is a very abundant population of *Camissonia crassifolia* near the northern limit of its distribution. In similar sandy situations, the species extends some 350 miles further south but its pollination was not studied over that range.

The locality in the Berkeley Hills of El Cerrito is a gentle, northwest-facing slope with a heavy clay-loam soil. In 1962 and 1963, when the site was intensively studied, it was one of the few vacant lots in an urban area and there was a concentration of plants of *Camissonia ovata* at the lower end. However, during the winter of 1963–64 this site was leveled in preparation for housing construction (pl. 5).

Most of the localities included in this and previous studies have been characterized by sandy soil which is optimum for most *Camissonia* species and provides ideal edaphic conditions for the construction of bee burrows. The only exception was *C. ovata* which along with its oligolectic bee utilizes the claylike soil of the western hill areas.

POLLEN PLANTS

The following species of *Camissonia* and *Oenothera* were included in this study:

Camissonia bistorta (Nutt. ex Torr. & Gray) Raven
C. boothii (Dougl.) Raven subsp. *decorticans* (Hook. & Arn.) Raven
C. californica (Nutt. ex Torr. & Gray) Raven
C. campestris (Greene) Raven subsp. *campestris*

C. campestris subsp. *obispoensis* Raven
C. cheiranthifolia (Hornem. ex Spreng.) Raimann subsp. *cheiranthifolia*
C. cheiranthifolia subsp. *suffructicosa* (S. Wats.) Raven
C. contorta (Dougl.) Kearney
C. crassifolia (Greene) Raven
C. graciliflora (Hook. & Arn.) Raven
C. micrantha (Hornem. ex Spreng.) Raven complex
C. ovata (Nutt. ex Torr. & Gray) Raven
C. sierrae Raven subsp. *sierrae*

Oenothera californica (S. Wats.) S. Wats.
Oe. deltoides Torr. & Frem. subsp. *cognata* (Jeps.) Klein
Oe. deltoides subsp. *howellii* (Munz) Klein
Oe. hookeri Torr. & Gray

The genus *Camissonia,* as presently recognized, was delimited and later monographed by Raven (1964, 1969). In the first three papers of the current series (Linsley, MacSwain, and Raven 1963*a,* 1963*b,* 1964) it was regarded as part of *Oenothera,* although always treated as a very distinct evolutionary line within the genus. In fact, evidence summarized in the two papers cited above demonstrates clearly that *Oenothera* and *Camissonia* are each more closely related to other genera of the family than they are to one another, and that their grouping was in fact an historical accident rather than a conscious biological decision. The relationships of all the genera in the tribe Onagreae of Onagraceae, to which both *Camissonia* and *Oenothera* belong, have been discussed in detail in these two papers, and are presented in summary form in the preceding paper of this series (MacSwain, Raven, and Thorp, 1973).

Of the species we discussed in the three first papers of this series as *Oenothera,* five—*Oe. albicaulis* Engelm., *Oe. caespitosa* Nutt., *Oe. deltoides* Torr. & Frem., *Oe. pallida* Lindl. [including *Oe. runcinata* (Engelm.) Munz and *Oe. trichocalyx* Nutt, as subspecies], and *Oe. primiveris* A. Gray—are currently regarded as members of the genus *Oenothera.* The remaining 12 species—*Camissonia boothii* (Dougl.) Raven (*Oenothera decorticans* Hook. & Arn.), *C. brevipes* (A. Gray) Raven, *C. californica* (Nutt. ex Torr. & Gray) Raven (*Oenothera leptocarpa* Greene), *C. campestris* (Greene) Raven, *C. cardiophylla* (Torr.) Raven, *C. claviformis* (Torr. & Frem.) Raven, *C. kernensis* (Munz) Raven subsp. *gilmanii* (Munz) Raven (*Oenothera dentata* Cav. var. *gilmanii* Munz), *C. k. kernensis, C. multijuga* (S. Wats.) Raven, *C. pallida* (Abrams) Raven subsp. *pallida* (*Oenothera abramsii* Macbride, *Oe. micrantha* Hornem. ex. Spreng. var. *pallida* (A. Nels) Munz), *C. refracta* (S. Wats.) *C. scapoidea* (Torr. & Gray) Raven, and *C. tanacetifolia* (Torr. & Gray) Raven—are now referred to the genus *Camissonia.* The synonymy just presented will allow the integration of our earlier papers with the current one.

Pollination Systems in Oenothera

Oenothera, with about 75 species in North and South America, consists primarily of species in which the hypanthium is more than 2 cm long. It is a long, slender, nectar-filled tube. The stigma is deeply divided into four slender lobes. The white or yellow flowers open near sunset, and they have a strong, often sweet odor. As might be expected from these characteristics, the outcrossing species of *Oenothera* are visited and pollinated primarily by hawkmoths (family Sphigidae). This has been elegantly documented by Gregory (1963–64) for several species, including especially *Oe. hookeri;* and in a number, including for example *Oe. primiveris,* which we have studied at a number of localities, bees are not even known to visit the flowers.

In others, bees do gather residual pollen from the flowers the morning after the flowers first open. Some early morning bees, for example *Andrena linsleyi* Timb., are oligolectic on evening-opening, basically hawkmoth-pollinated species of *Oenothera* (Linsley, MacSwain, and Raven, 1963*a*). Exceptionally, bees—for instance *Xylocopa* (Gregory, 1963–64)—may utilize the flowers of *Oenothera* for pollen in the early evening and, by biting through the base of the hypanthium, can also obtain nectar from these flowers. However, the evidence is overwhelming that bees play little role in the pollination of *Oenothera* with a few exceptions to be mentioned below.

These exceptions include *Oenothera* subg. *Kneiffia* of the eastern United States and adjacent Canada. All of the members of this group have bright yellow flowers that open in the morning. Bee-pollination is predominant, although little studied, and moths do not visit the flowers of *Kneiffia,* which are odorless and have relatively short hypanthia.

Bee pollination has also originated independently at least three times in *Oenothera* subg. *Hartmannia,* which also includes moth-pollinated, evening-opening species. One of the bee-pollinated groups is yellow-flowered, and ranges from central Mexico to Peru and Bolivia (*Oenothera multicaulis* complex). From this group have been derived the only hummingbird-pollinated populations in the genus *Oenothera,* marked by a large red splotch at the base of each petal. These occur at relatively high elevations from central Mexico to Costa Rica, and again in southern Columbia and Ecuador. Two other groups of *Oenothera* subg. *Hartmannia* which have flowers that open in the morning include the only purple-flowered species of the genus. They are likewise bee-pollinated.

Pollination Systems in Oenothera

Camissonia, which consists of 60 species of western North America and one of temperate western South America (Raven, 1969), consists primarily of species that have hypanthia less than 8 mm long and bright yellow odorless flowers that open near sunrise. All species of the genus have a capitate stigma.

From this basic group have been derived one group of two closely related species—sect. *Lignothera* (including *C. cardiophylla*)—in which the yellow flowers open near sunset and the hypanthia are 4.5 to 40 mm long. As might have been suspected, these species are visited by and primarily pollinated by hawkmoths.

As pointed out by Raven (1969: 174), the outcrossing species of the entirely white-flowered, evening-opening sect. *Eremothera* (*C. boothii* and *C. refracta*) and the white-flowered and afternoon-opening yellow-flowered populations of *C. claviformis* (sect. *Chylismia*) are primarily pollinated by medium-sized moths (mainly Noctuidae and Geometridae). In each group, however, oligolectic bees have become associated with the plants; thus, the relationship of *Andrena vespertina* Linsley & MacSwain with *C. boothii* parallels that of *A. rozeni* Linsley & MacSwain with *C. claviformis.*

We reported earlier (Linsley, MacSwain, and Raven, 1963*a,* 1964) that *Andrena rozeni* Linsley and MacSwain may be the chief pollinator of at least the California white-flowered subspecies of *C. claviformis.* It appears likely that in these populations, the highly oligolectic and abundant bee may be essentially taking over as principal pollinator of the plants, displacing the medium-sized moths that are in an evolutionary sense responsible for their shift to evening opening and white flower color. The preeminence of the bees may account for the time of opening of these plants, which is some two hours before sunset—highly unusual in a moth-pollinated plant.

Two species of sect. *Chylismia, C. megalantha* (Munz) Raven, a local endemic of southern Nevada, and *C. heterochroma* (S. Wats.) Raven, its more widespread but highly colonial autogamous derivative, are the only purple-flowered species of *Camissonia.* The former is visited abundantly and pollinated by the polylectic bees *Centris rhodopus* Ckll. and *Anthophora urbana* Cress. soon after its flowers open near sunrise (Raven, ms).

Finally, *C. pleurosperma* (S. Wats.) Raven (of the monotypic sect. *Chylismiella*) has rather small white petals that are yellow near the base and flowers that open in the morning. This pattern of petal coloration is found elsewhere in the tribe Onagreae only in the genus *Gayophytum,* where pollination of outcrossing species by flower-visiting flies predominates and bees are occasional (Raven and Thorp, ms.); but *C. pterosperma* is largely if not entirely autogamous.

Oenothera in Cismontane California and Baja California

Oenothera is rather poorly represented in the area of the present study, being more diverse as to species and abundant as to individuals in the Great Basin, Great Plains, and Mexico. Of the three species we encountered in the course of our field work two—*Oe. californica* and *Oe. deltoides*—are white-flowered and the third, *Oe. hookeri,* is yellow-flowered. All have flowers with a strong odor and hypanthia 2–4 cm long, and each is predominantly pollinated by hawkmoths, both within the area of the present study and outside of it. All three species occur primarily in regions of sandy soil.

Oenothera hookeri, which is a polytypic species that ranges widely in the mountains of the western United States and adjacent Mexico (fig. 1), has been studied in great detail from the point of view of its pollination biology (Gregory, 1963–64). It is a tall clumped biennial or perennial, often found in relatively moist sites, that bears its yellow flowers on the upper third of wandlike stems that may reach 2.5 meters in height (pl. 3). Its yellow flowers are 4.5–10 cm across and open near sunset. *Oenothera hookeri* is pollinated almost exclusively by hawkmoths, although in certain situations various species of bees gather residual pollen from its flowers. It is a member of subg. *Oenothera.*

The other two species of *Oenothera* are white-flowered and are closely related members of subg. *Anogra. Oenothera californica* is a rhizomatous perennial that occurs near the coast from San Luis Obispo County south to southern California

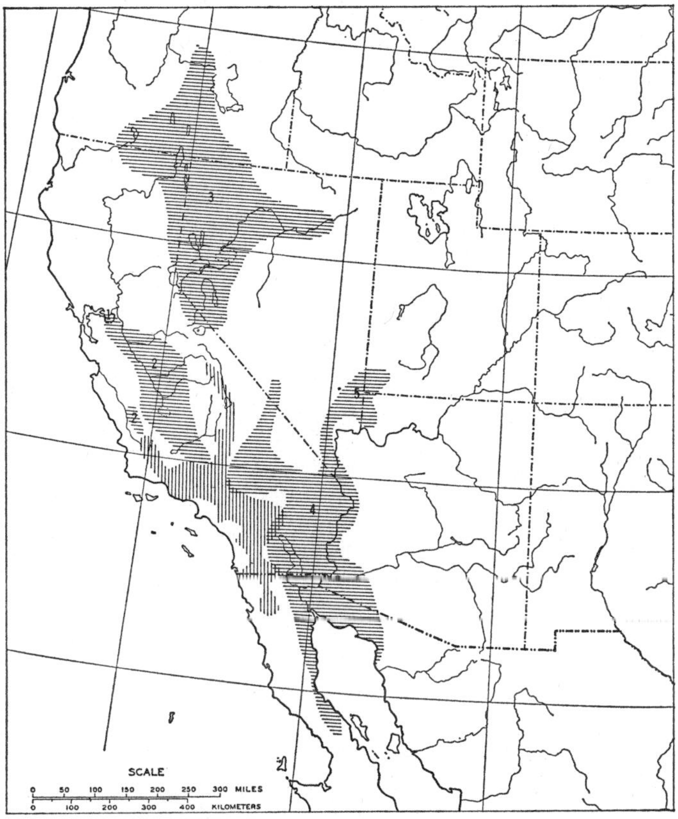

Fig. 1. Distribution of *Oenothera deltoides* subspecies (1–5) horizontal hatching (1—*Oe. d. howellii.* 2—*Oe. d. cognata;* 3—*Oe d. piperi;* 4—*Oe. d. deltoides;* 5—*Oe. d. ambigua*)*; and Oenothera californica,* vertical hatching.

and adjacent Baja California, thence eastward onto the borders of the desert (fig. 2). Its flowers are 3.5–7.5 cm across and, like *Oe. hookeri,* it is rarely visited by bees, as we found in our earlier studies (Linsley, MacSwain, and Raven, 1964).

Oenothera deltoides cognata is a robust, well-branched annual that may be up to 5 dm tall. It occurs in sandy places in San Joaquin Valley (fig. 2) and may overwinter if conditions are particularly favorable. Like *Oe. hookeri* but unlike the subspecies of *Oe. deltoides* we studied earlier on the Colorado and Mojave Deserts and in the Great Basin, *Oe. d. cognata* consists entirely of self-compatible individuals. Its white flowers are 4.5–8 cm across (pl. 6), and they are visited and primarily pollinated by hawkmoths and the nocturnal bees, *Lasioglossum* (*Evylaeus*) sp. near *galpinsiae.*

Oenothera d. howellii is a local endemic of the sand dunes along the San Joaquin River at Antioch, Contra Costa County, California (fig. 2), a habitat that has essentially vanished under the influence of urban and industrial use. It differs primarily from *Oe. d. cognata* in being somewhat rhizomatous and therefore perennial, and in its self-incompatibility. The plants are densely covered with

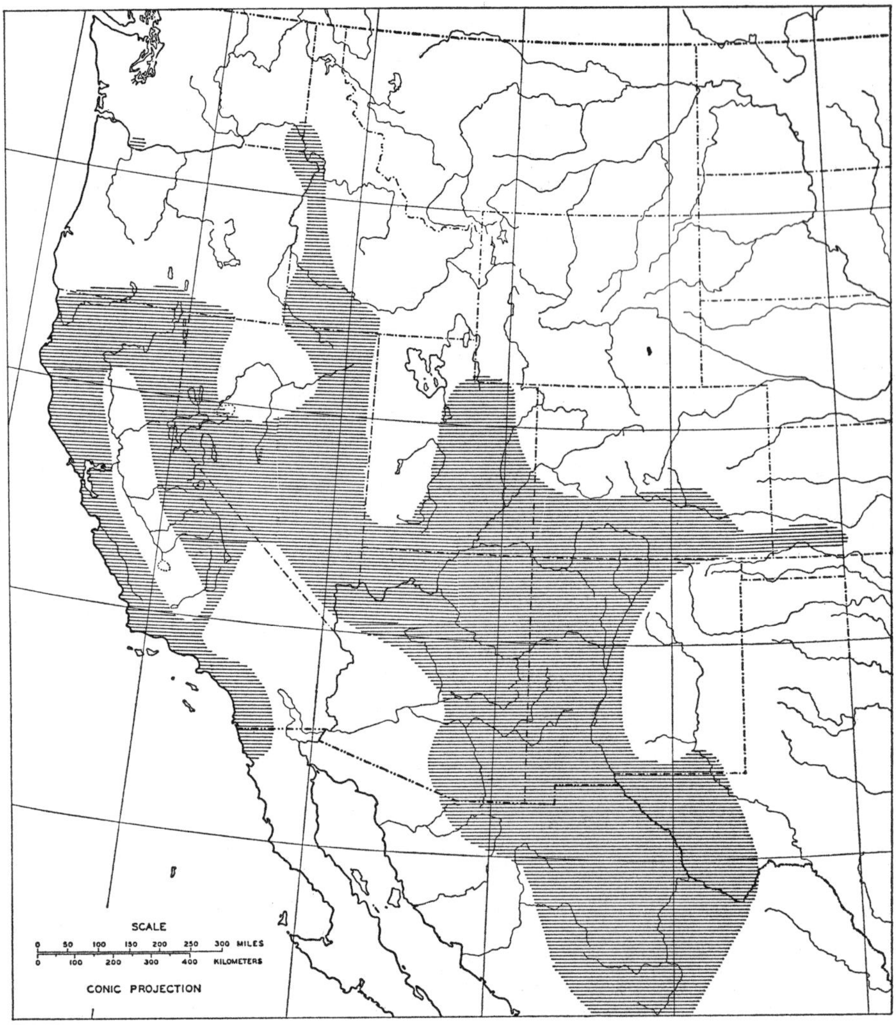

Fig. 2. Distribution of *Oenothera hookeri*.

short gray hairs, whereas those of *Oe. d. cognata* are characterized by long, spreading hairs on the upper parts. A few populations of *Oe. d. cognata* in the vicinity of Antioch appear to have been modified by introgression from *Oe. d. howellii*, but it is in general a well-marked local taxon. In terms of its floral biology, it is equivalent to the other subspecies of *Oe. deltoides:* diurnal bees, while sometimes abundant on its flowers, play a very minor role in its pollination. However, the crepuscular bee *Lasioglossum* (*E.*) sp. near *galpinsiae* may be an important pollinator.

Camissonia in Cismontane California and Baja California

Whereas *Oenothera* is relatively poorly represented in the area under consideration, it is perhaps the major center of evolution for *Camissonia*. The nine sections of this genus are sharply distinct, and it is convenient to review each of them in terms of their general characteristics and floral biology.

Camissonia sect. *Eulobus* consists of three species, two of them self-incompatible, that are restricted to Baja California, and one more widespread, largely autogamous species, *C. californica* (*Oenothera leptocarpa*), that ranges north to central California and east to Arizona and Sonora. All four species of this section have a fine red flecking on the lower third of the petals, and this section is likewise non-ultraviolet reflective, in contrast to the rest of the petals, which are strongly ultraviolet reflective. We have studied the pollination biology of only one of the Baja California endemics, *C. crassifolia* (fig. 3), and that only in the northern part of its range, but the results obtained with this shrubby, self-incompatible species (which can also flower as an annual) were highly interesting. In *C. crassifolia,* which may be the most primitive species of the genus, the flowers are 1.5–3.5 cm across. In the largely self-pollinating *C. californica,* which is rarely visited by bees (Linsley, MacSwain, and Raven, 1963*a*) they are not much smaller: 1.2–3 cm across. It is most unfortunate that we were unable to study the pollination biology of the second self-incompatible species of the section, *C. sceptrostigma* (T. S. Brandegee) Raven, an annual of central and southern Baja California.

The pollination biology of *Camissonia* sect. *Chylismia* has been summarized above and is treated in detail in our earlier papers (Linsley, MacSwain, and Raven, 1963*a,* 1963*b,* 1964). This group of 13 species is confined to the deserts and Great Basin of the western United States and adjacent Mexico, and does not enter the region of the present study. The same is true of the small sections *Lignothera* and *Chylismiella,* discussed above, and *Nematocaulis,* which comprises two closely related autogamous species of the northern Great Basin. The pollination biology of the white-flowered sect. *Eremothera,* of which *C. boothii* subsp. *decorticans* is the only representative in cismontane California (pl. 6, fig. 4), has likewise been discussed above. Other members of this group are confined to the deserts and Great Basin.

The three remaining sections of *Camissonia* contain the species with which we were most concerned during the course of the present study. All species of these groups have bright yellow flowers that open in the morning. Of the six species sect. *Taraxia,* two—the perennial, self-incompatible *C. ovata* and the annual, autogamous *C. graciliflora*—occur in cismontane California and were studied in detail. The other self-incompatible species of this section, distantly related to the species considered here, is *C. tanacetifolia* of the Great Basin. We earlier (Linsley, MacSwain, and Raven, 1963*b*) reported upon its most interesting floral biology in detail.

Camissonia ovata is a clumped perennial of clay soils near the coast from southern Oregon to northern San Luis Obispo County, California (fig. 5). Its largely ultraviolet reflective bright yellow flowers are borne among the basal leaves and elevated above them by means of a long, slender sterile projection at the tip of the ovary. To the naked eye, they appear unspotted, but in fact they have a large, non-ultraviolet reflective "target" occupying the center of the flower. The flowers are mostly 2–5 cm across.

In the autogamous (frequently cleistogamous) annual *C. graciliflora,* which is

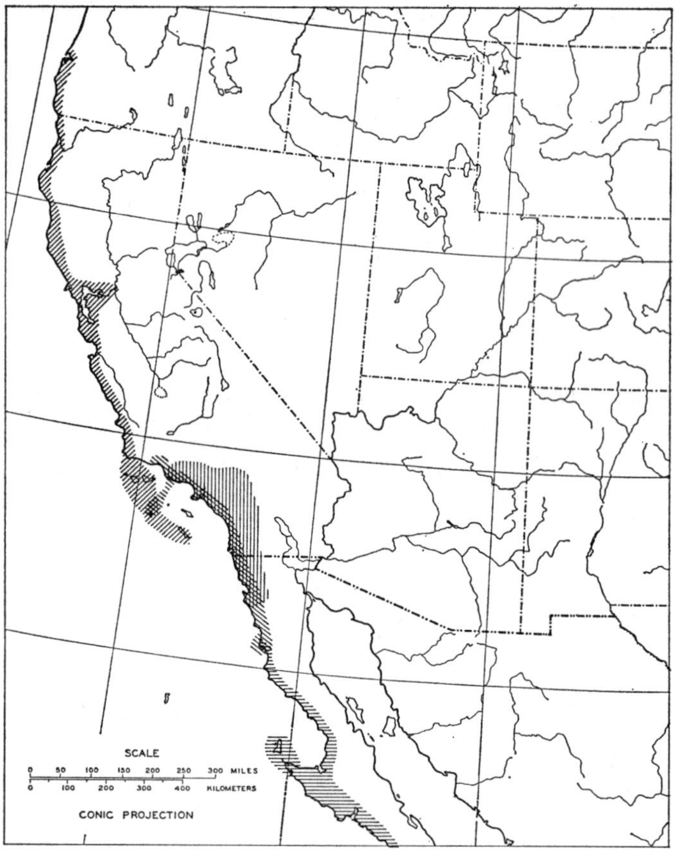

Fig. 3. Distribution of *Camissonia cheiranthifolia cheiranthifolia,* 45° hatching; *Camissonia c. suffruticosa,* 315° hatching; *Camissonia bistorta,* vertical hatching; and *Camissonia crassifolia,* horizontal hatching.

widespread from southern Oregon throughout cismontane California to northern Baja California, the clear yellow flowers are only 1–3.5 cm across, and are rarely visited by bees.

Camissonia sect. *Holostigma* is virtually endemic to the area under consideration here. Of the 14 species included in this group, 12—including *C. pallida,* which we studied earlier (Linsley, MacSwain, and Raven, 1963*a*, 1964)—are autogamous and rarely visited by bees, while two—*C. cheiranthifolia* and *C. bistorta*—are outcrossers that were very important elements in our present study.

Camissonia cheiranthifolia, the only perennial species of the section, is strictly confined to sand dunes along the ocean, where it occurs from Coos County, Oregon, to northern Baja California (fig. 3). From Point Conception, Santa Barbara County, California and the Channel Islands, northward, the species is represented by subsp. *cheiranthifolia,* which is strictly herbaceous and has flowers only 12–25 mm across that usually do not have red dots near the base of the petals (pl. 6). This subspecies is uniformly self-compatible and regularly self-pollinated, but

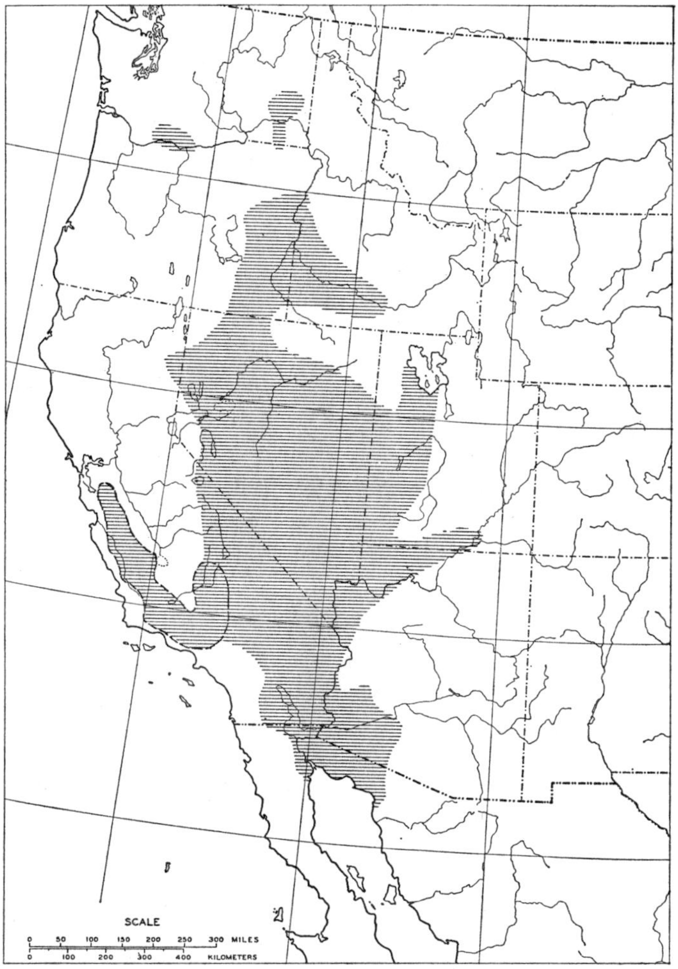

Fig. 4. Distribution of *Camissonia boothii* with the cismontane California subspecies *C. b. decorticans* outlined by the dashed line.

still exhibits a high degree of outcrossing at those relatively few localities where it is associated with large populations of (oligolectic) bees. At the few inland localities for the species, on sand dunes along the Sacramento River from Antioch upstream for perhaps 10 miles, bees were likewise abundant. Most of the colonies of this subspecies occur in places at which heavy morning fog, not conducive to the activities of bees, is characteristic, its prevalent autogamy is therefore presumably highly adaptive.

From the vicinity of Santa Barbara southward, where morning fogs are not nearly so prevalent as farther north, the dunes are occupied by the shrubby *C. c. suffruticosa* (S. Wats.) Raven, in which the flowers are usually 25–45 mm across and each petal normally has one or two large reddish brown dots near the base. Although many of the populations of this subspecies are self-compatible,

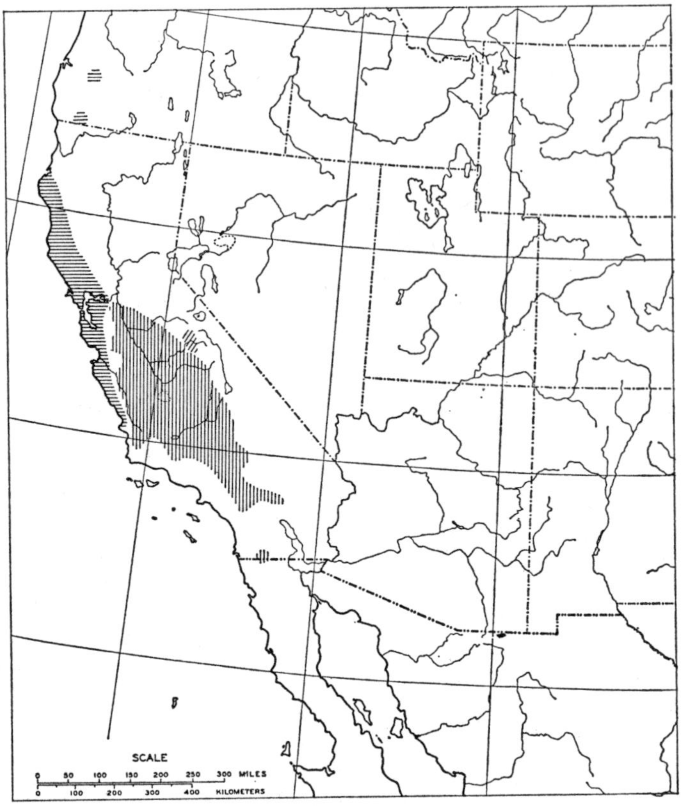

Fig. 5. Distribution of *Camissonia ovata,* horizontal hatching; *Camissonia campestris,* vertical hatching; and *Camissonia sierrae sierrae,* 45° hatching.

some—in San Diego County, California—consist of self-incompatible individuals. The stigma is elevated well above the anthers and all populations, whether self-incompatible or not, are highly outcrossing and abundantly visited by bees.

Ranging from the coast inland in the area occupied by *C. c. suffruticosa* is the annual *C. bistorta* (fig. 3) which is entirely self-incompatible and much frequented by bees. *Camissonia bistorta* often forms large colonies in sandy soil, as in washes. It is florally almost identical with *C. c. suffruticosa,* although its flowers rarely exceed 30 mm in diameter. Where the two come into contact in San Diego County and northern Baja California, as near the mouth of rivers that reach the sea, hybrid swarms are frequent. It is obvious that the separation between these two entities is primarily ecological.

The smaller-flowered species of sect. *Holostigma* are highly autogamous. Their flowers are generally less than 15 mm across, and they are rarely visited by bees except where they are growing mixed with other, larger-flowered species or under conditions of pollen stress (Linsley, MacSwain, and Raven, 1964). They are here grouped as the *Camissonia micrantha* complex.

All of the species of sect. *Camissonia* are slender, wiry-stemmed annuals restricted to sandy soils. Two—*C. campestris* and *C. kernensis*—are self-incompatible. These were considered during our study on the Mojave Desert (Linsley, MacSwain, and Raven, 1964). One additional taxon—*C. sierrae*—is self-compatible but outcrossing. The remaining nine species are autogamous.

Of the two species that occur in the region of the present study, *C. campestris* ranges throughout the San Joaquin Valley and bordering hills, whereas *C. sierrae* subsp. *sierrae* is confined to the central Sierra Nevada of Mariposa and Madera counties (pl. 4, fig. 5). In the former, the flowers are generally 12—30 mm across. Each petal usually has two large reddish brown dots near the base (pl. 6). In the latter, the flowers are generally 9–15 mm across, and the petals are likewise red-dotted. The two species do not occur sympatrically. *Camissonia campestris obispoensis* is a geographical race of coastal northern San Luis Obispo County and adjacent Monterey County (fig. 5) which differs from *C. c. campestris* largely in its broader, coarsely dentate leaves.

In the autogamous species of sect. *Camissonia,* the flowers are mostly less than 10 mm across. Visits by bees are almost unknown, except under the conditions noted for autogamous species of sect. *Holostigma.* The only South American species of the genus *C. dentata* Cav. of southern Peru, Chile, and Argentina, is one of these autogamous species which presumably reached South America by long-distance dispersal across the tropics.

BEE VISITORS OF CAMISSONIA AND OENOTHERA

Andrena

Eighteen species of *Andrena* were found to be oligolectic on *Camissonia* in areas covered in the present study. These belong to two subgenera: nine of these are *Onagrandrena* and nine are *Diandrena.*

Field observations at the site east of Bakersfield were made from late February through the end of April between 1958 and 1962 but the activity span of *Andrena* at this locality was from the first week of March until mid-April or approximately five to six weeks. At other localities these bees were still active toward the end of May but the start of their season was not determined. We have seen collection records of *A. oenotherae* as late as 20 July. The overwintering stage of one species of each subgenus was determined to be the adult with transformation to this stage occurring during the summer months. Our evidence suggests that many of these bees are capable of waiting over an unfavorable season without emergence.

The following key should help to distinguish the two subgenera.

KEY TO SUBGENERA OF ANDRENA OLIGOLECTIC ON CAMISSONIA

1. Integument black; forewings with three submarginal cells; usually large species . . . *Onagrandrena*

   Integument metallic blue or green, if black then with metallic bluish reflections on metasoma; forewings with two submarginal cells; usually small species . . . . *Diandrena*

Andrena (Diandrena)

Nine species of *Andrena* (*Diandrena*)—*A. anatolis* Linsley & MacSwain, *A. apasta* Linsley & MacSwain, *A. chalybaea* Cresson, *A. cyanosoma* (Cockerell), *A. eothina* Linsley & MacSwain, *A. foxii* Cockerell, *A. macswaini* Linsley, *A. parachalybea* Viereck, and *A. sperryi* (Cockerell)— are oligolectic on yellow-flowered, morning-opening camissonias in cismontane California and northwestern Baja California, Mexico. Four of these—*A. apasta, A. foxii, A. parachalybea,* and *A. sperryi*—also occur sparingly on the western margin of the Mojave Desert and were treated earlier (Linsley, MacSwain, and Raven, 1964). Six other species, *A. agoseridis* Thorp, *A. chlorosoma* Linsley & MacSwain, *A. malacothricidis* Thorp, *A. olivacea* Viereck, and *A. subchalybea* Viereck, which gather pollen from Compositae and *A. cuneilabris* Viereck, which gathers pollen from *Ranunculus,* occasionally visit *Camissonia* for nectar (Thorp, 1969), but will not be discussed here. The principal field observations on *Diandrena* associated with *Camissonia* were conducted between 1959 and 1964. Since this subgenus has been revised recently by Thorp (1969), the species are only treated briefly below.

Because of the superficial similarity of these bees, keys are offered to help distinguish species of *Diandrena* whose females collect pollen from *Camissonia.*

KEY TO DIANDRENA OLIGOLECTIC ON CAMISSONIA IN CISMONTANE CALIFORNIA AND BAJA CALIFORNIA

Females

1 Head and mesosoma metallic blue or green, pubescence predominantly pale 2
Head and mesosoma black, pubescence black; length of body 11–12 mm, anterior wing 8–8.5 mm .... *foxii*

2 (1) Metasomal tergites subopaque or opaque, interspaces between punctures sculptured 3
Metasomal tergites shining, interspaces between punctures smooth; length of body 6.5–7 mm, anterior wing 5 mm .... *sperryi*

3 (2) Hind wing with jugal lobe broad and with incision between it and vannal lobe 4
Hind wing with jugal lobe narrow and without incision between it and vannal lobe 7

4 (3) Mesoscutum rugoso-punctate, punctures contiguous 5
Mesoscutum not rugoso-punctate, punctures separated by one diameter; length of body 8–10 mm, anterior wing 7–8.5 mm. .... *apasta*

5 (4) Metasomal tergites moderately shining, punctures distinct, larger species 6
Metasomal tergites opaque, punctures small, shallow, obscure; length of body 7–9 mm, anterior wing 5.5–6.5 mm .... *cyanosoma*

6 (5) Facial pubescence predominantly black; integument blue-green; metasoma with punctures of disc of second tergite large, separated by two diameters; length of body 10.5–12 mm, anterior wing 8.5–9.5 mm .... *chalybaea*
Facial pubescence predominantly white; integument blue; metasoma with punctures of disc of second tergite small, separated by one diameter or less; length of body 10–11 mm; anterior wing 7.5–8.5 mm .... *parachalybea*

7 (3) Propodeal enclosure finely rugulose basally, transversely tesselate apically; smaller species. 8
Propodeal enclosure irregularly rugulose with longitudinal median carinula; length of body 9–9.5 mm, anterior wing 6.5 mm .... *macswaini*

8 (7) Labral process deeply emarginate apically; integument blue with green reflections; length of body 7.5–8 mm, anterior wing 5.5–6 mm .... *eothina*
Labral process feebly emarginate apically; integument green; length of body 6–7 mm, anterior wing 5–5.5 mm .... *anatolis*

Males

1 Head and mesosoma metallic blue or green 2
Head and mesosoma black; length of body 7.5–10 mm, anterior wing 7–7.5 mm ..... *foxii*

2 (1) Metasoma with interspaces between punctures sculptured 3
Metasoma with interspaces between punctures smooth; length of body 5.5–6 mm, anterior wing 4.4–5 mm .... *sperryi*

3 (2) Jugal lobe of hind wing broad and with an incision between it and vannal lobe 4
Jugal lobe of hind wing narrow and without an incision between it and vannal lobe 7

4 (3) Mandibles long, decussate; mesoscutum rugoso-punctate with punctures usually contiguous. 5
Mandibles short, apposite; mesoscutum with punctures usually separated by one diameter. 6
5 (4) Blue-green; labral process weakly emarginate apically; disc of second metasomal tergite with interspaces between punctures distinctly imbricate; length of body 9.5–10 mm, anterior wing 8.5–9 mm .... *chalybaea*
Blue; labral process distinctly emarginate apically; disc of second metasomal tergite with interspaces between punctures almost smooth with at most faintly indicated imbrications; length of body 8–8.5 mm, anterior wing 6.5–7 mm .... *parachalybea*
6 (4) Gray-blue; labral process broadly rounded and weakly emarginate apically; metasoma with interspaces between punctures coarsely, closely reticulate; apex of eighth sternite emarginate; length of body 7–8 mm, anterior wing 5–6 mm .... *cyanosoma*
Olive-green; labral process deeply emarginate apically with teeth pointed; metasoma with interspaces between punctures imbricate; apex of eighth sternite arcuate, with a small median lobe; length of body 7–8 mm, anterior wing 6–6.5 mm .... *apasta*
7 (3) Propodeal enclosure longitudinally, sparsely rugulose basally, transversely tesselate apically; mandibles moderately long, decussate........ 8
Propodeal enclosure obliquely, densely rugulose with a median longitudinal carinula; mandibles short, apposite; length of body 7.5–8 mm, anterior wing 6–6.5 mm.. *macswaini*
8 (7) Green; labral process feebly emarginate apically; length of body 6–6.5 mm, anterior wing 4.5–5 mm .... *anatolis*
Blue with greenish reflections on head; labral process deeply emarginate apically; length of body 6–7 mm, anterior wing 5–5.5 mm .... *eothina*

Andrena (*Diandrena*) *chalybaea* (Cresson)

Andrena chalybaea, the type of the subgenus *Diandrena,* is known only from the Coast Ranges of central California from Mendocino County south to San Luis Obispo County (fig. 6). Its only known pollen source is *Camissonia ovata.* Our observations are based on 512 females and 6 males. Both males and females are on the wing from early March through late April. Additional records (Thorp, 1969) show that both sexes are active by mid-February.

Males of *A. chalybaea* spend most of their active hours cruising over areas of exposed soil where the females are nesting or over the host plants. In the latter case it is difficult to tell whether the attraction is to the plants since the females commonly nest in areas of sparse vegetation in among the plants of *Camissonia ovata.* The paucity of records of males as compared with females taken in association with *C. ovata,* the females being almost 15 times as abundant as the males at these plants, indicates that the males search for the females principally in the vicinity of their nests and obtain nectar from other flowers. Daily activity for males begins about two hours after the initiation of female pollen-collecting activity. Thus male activity does not greatly overlap with the period of floral availability of *C. ovata.* The records of males and females on various nectar sources which are utilized later in the morning after the flowers of *C. ovata* have closed are closer to a one-to-one ratio. These nectar sources include: *Ranunculus californicus, Blennosperma bakeri, Brassica* sp., and *Cryptantha* sp.

Mating attempts have been observed at the flowers of *Ranunculus* and with pollen-laden females at flowers of *Camissonia.* However, the males were never successful in inserting their genitalia. Presumably then the nest areas of the females are the sites of most of the successful mating activity, especially in the early season when the females are first emerging from overwintering, just as is

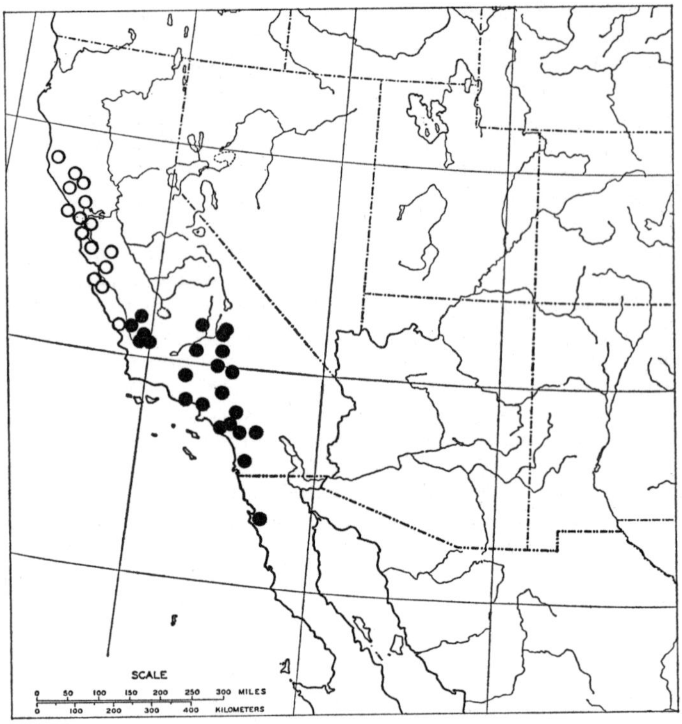

Fig. 6. Distribution of *Andrena* (*Diandrena*) *chalybaea*, circles; and *Andrena* (*D.*) *parachalybea*, dots.

the case for *A. foxii* (see below). Males spend the night in the soil either in old female burrows or in shallow burrows they excavate themselves.

Burrows of *A. chalybaea* are constructed in the heavy clay-loam soils which are typical of the area where the bee occurs. They are most commonly encountered in flat areas devoid of vegetation or where the vegetation is sparse. The entrance is surrounded by a concentric tumulus. The vertical shaft penetrates the soil to a depth of 7.5 to 10 cm. A lateral shaft of 5 to 10 cm leads to the terminal cell which is slightly inclined below the horizontal. Presumably there is more than one cell associated with a single vertical entrance, but only the lateral leading to the cell currently under construction and provisioning is open, all preceding laterals having been plugged. This is indicated by the fact that other cells of the proper depth and distance from the vertical shaft are often unearthed during the excavation of a nest. The cell is highly polished and lined with a thin coat of waterproof material over all its inner surface except the cell cap.

Females of this species collect pollen exclusively from *Camissonia ovata* (pl. 1). Nectar requirements are supplemented by visits to *Ranunculus californicus* and *Blennosperma bakeri*, but usually only after the daily foraging activities on *C. ovata* are drawing to a close. About 14 times as many females were taken on *C. ovata* as on any other plant. Daily pollen foraging is restricted to the early morning hours for a period of approximately two to four hours beginning shortly after the sun appears on the site. The time females spend foraging on individual flowers is inversely related to the time which the flowers have been open for the day (Thorp, 1969: 12). Females orient across the center of the flower contacting both stigma and stamens when gathering pollen.

Pollen provisions in the cells are initially dry. They are eventually worked into a spheroidal mass through the addition of small quantities of nectar. The final mass is arched upwards and flattened on the bottom. Its equatorial diameter is about 5 to 6 mm and its depth 4 to 5 mm. The egg is laid on top of the provisions. The larvae have completed their feeding and defecation by the end of the spring and the post-defecating larva remains quiescent throughout the summer. The feces are deposited at the bottom of the cell in a lens-shaped mass. No cocoon is spun. Pupation occurs in September or October with the males pupating before the females. The bees overwinter as adults in their cells and emerge the following February. There is no evidence that any individuals of this species remain in their cells for more than one year.

The principal parasite of this species is an unnamed species of *Nomada*. Its life history and behavior have been described by Thorp (1969). A few mite hypopi have been found on the adults of *A. chalybaea*, but their relationship to the bees is unknown. European earwigs, *Forficula auricularia L.*, are occasionally found in the open vertical shafts of old nests of the bee, but no damage to the cells was observed.

This species is most numerous where populations of its pollen source are most abundant. The bees are less commonly encountered as one moves northward from the San Francisco Bay region. The most northerly records are well south of the northern limits of its pollen plant. At the southern end of the range sizeable populations of the bee have been discovered at Cambria, San Luis Obispo County, which is the southern limit of the distribution of *C. ovata*.

Andrena (*Diandrena*) *parachalybea* Viereck

Andrena parachalybea, the closest relative of *A. chalybaea*, occurs in southwestern California and northwestern Baja California (fig. 6). It is found along the coast in northern Baja and extreme southern California, extending eastward into the western edge of the Mojave Desert and northward into the interior valleys. Eastern populations of this species have been previously treated (Linsley, MacSwain, and Raven, 1964). Females collect pollen from several species of *Camissonia*, principally *C. campestris* in the interior areas, *C. bistorta* both interiorly and coastally, and *C. cheiranthifolia suffruticosa* in the coastal areas. It has been reported to collect pollen from *C. pallida pallida* (Abrams) Raven (as *Oenothera abramsii* MacBride) by Linsley, MacSwain, and Raven (1964) in the Mojave Desert. Our observations are based on 210 females and 4 males. Both males and females are generally on the wing from late March to early May. However, Thorp (1969) has shown that both sexes may be active in late February.

Males of *A. parachalybea* have not been found on the plants which serve as sources of pollen for the females, but have been taken principally at flowers which they visit for nectar. Nectar records for the males include: *Haplopappus, Lasthenia,* and *Descurania.* Nectar records for the females include: *Salix, Cryptantha, Erodium,* and *Layia.*

Females of this species collect pollen only in the early morning, beginning their activity within an hour after the sun is on the site and continuing into midmorning.

This bee is most abundant and a dominant *Diandrena* in the competitive complexes near Sage and at Point Dume (table 9). Its abundance decreases northward into the valleys of the inner Coast Ranges and into the southern end of the Central Valley (tables 1, 4, 6, 8, 9) and southward into Baja California.

Andrena (*Diandrena*) *foxii* Cockerell

Andrena foxii has been assigned to the Chalybaea group (Thorp, 1969), but it is not as closely related to *A. chalybaea* and *A. parachalybea* as they are to each other. It is known only from central and southern California from Merced County to Los Angeles County, extending sporadically eastward into the western and southern margins of the Mojave Desert (fig. 7). The rare and small populations of this bee in the Mojave Desert have been discussed (Linsley, MacSwain, and Raven, 1964). Females collect pollen exclusively from *Camissonia campestris.* Our observations are based on 640 females and 96 males. Both sexes are on the wing from early March to early May.

Males of *A. foxii* have been most commonly taken in association with flowers of *C. campestris.* The disparity of male versus female records for this species may be because the males are not completely dependent on the flowers of *Camissonia* for nectar, but are principally searching for females in these areas. Males of this species set up small "territories" within the area of *Camissonia* bloom (pl. 1). They occasionally sit on the flowers, but more frequently sit on the sandy soil amongst the flowers, darting out at moving objects encroaching on their areas and returning to their former areas when their investigations are completed. They also cruise defined areas of flowers and may come to rest at any of several open spots in the cruising area. Although the female burrows are also interspersed among the *Camissonia* plants, the male "resting positions" are not necessarily associated with these nests. Thorp (1969) records male nectar visits at the flowers of "Compositae," *Coreopsis,* and *Cryptantha.* One male was also taken at the flowers of *Lupinus* although these do not produce nectar. Males of *A. foxii* mate with the females as soon as the latter emerge from the ground. In one case copulation occurred immediately after a pair were placed in a cyanide jar. Many unsuccessful mating attempts have been observed with pollen-collecting, presumably mated, females which entered male "territories."

Burrows of *A. foxii* are constructed in light sandy soils. The entrances are most commonly encountered in flat areas of sparse vegetation. The entrance is surrounded by a concentric tumulus. Nests were most frequently found adjacent to plants with a basal rosette of leaves (e.g., *Erodium*). Plaster casts (pl. 2) revealed that the vertical shaft of the L-shaped burrows penetrates the soil to a depth of 19 to 32 cm (average of 4 nests = 25 cm). The lateral shaft leading to the cell ranged from 7.5 cm to 17.5 cm (average of 4 nests = 11.5 cm). The terminal cell

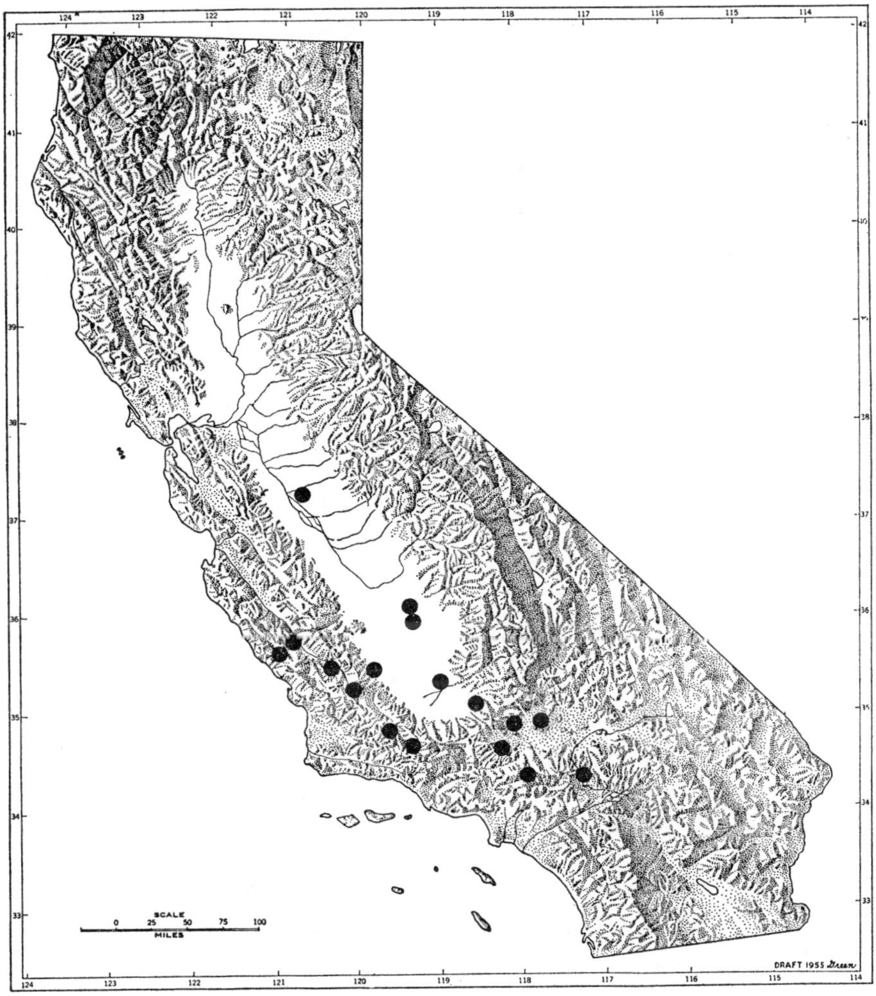

Fig. 7. Distribution of *Andrena* (*Diandrena*) *foxii.*

of the lateral is inclined slightly below the horizontal. The position and orientation of additional cells encountered in the excavation of these nests indicate that this species constructs several laterals sequentially from the same vertical shaft, with each lateral being filled with soil upon completion of the terminal cell. The inner walls of the cell are coated with a very thin transparent layer of waterproof material, which does not extend to the inner face of the cell cap. The inner surface of the cell cap exhibits a spiral arrangement of the soil particles.

Females collect pollen exclusively from *Camissonia campestris* (pl. 1). Nectar requirements are supplemented by visits to "Compositae" and *Hemizonia.* Thorp (1969) also records female nectar visits to *Coreopsis, Malacothrix,* and *Cryptantha,* as well as one female being taken at the nonnectar producing flowers of *Lupinus.* Females began to forage for pollen within an hour after sunrise, and before other species of *Diandrena,* which are oligolectic on *C. campestris,* become active (tables 1, 3, 4, 6, 7, 8). They collect pollen for a period of about two hours. During the last half of their diurnal activity period the number of flowers visited per minute increases while the number of flowers from which pollen is collected decreases (table 2). They cease activity while smaller females and other species of *Diandrena* continue to collect residual pollen from *Camissonia.* Females of *A. foxii* orient across the center of the flower contacting the stigma and stamens when gathering pollen and do not move in a circular path around the flower. In taking nectar they alight on the petals on the lower side of the flower and push their mouthparts into the centrally located nectary.

The pollen ball and the position of the egg in the cells of *A. foxii* are essentially like those described above for *A. chalybaea.*

This species is most abundant in the southern end of the Central Valley and in the valleys of the inner Coast Ranges. In these areas it is usually a dominant form, its females being the largest, darkest, and earliest flying of the *Diandrena.* It also occurs sparingly, in low population densities, along the eastern margin of its distribution.

Andrena (*Diandrena*) *sperryi* (Cockerell)

Andrena sperryi, one of the smallest of the *Diandrena,* occurs in central and southern California, from Merced County to Los Angeles County, extending into the western and southern edges of the Mojave Desert (fig. 8). In cismontane California the females collect pollen only from *Camissonia campestris.* In the Mojave Desert it also collects pollen from *C. pallida* (Linsley, MacSwain, and Raven, 1964) (reported as *Oenothera abramsii*). Our records are based on 42 females and 10 males. The females usually fly from mid-March to early May and the males from early March to mid-April. However, males have also been found in early May (Thorp, 1969).

Eight males of *A. sperryi* have been taken in association with the female pollen source, *C. campestris,* another male was taken on *C. contorta.* In addition Thorp (1969) records three males on *Malacothrix.* The paucity of male records as compared with females in association with *C. campestris,* the females being almost six times as abundant as the males at these flowers, indicates that the males may search for the females principally in the vicinity of their nests.

Females collect pollen principally from *Camissonia campestris.* Their nectar requirements are supplemented by visits to Compositae. Thorp (1969) records female nectar visits to *Coreopsis, Malacothrix,* and *Anisocoma.* Females of *A. sperryi* are the smallest of the *Diandrena* which are oligolectic on *Camissonia* and they initiate their daily pollen foraging late in the morning after the other species have passed their peak foraging activities and only about an hour before the flowers close for the day (tables 6, 8). They appear to be principally scavengers of residual pollen remaining after the other species have nearly completed their pollen collecting. Due to their small body size, they rarely come into contact with the stigma while scraping pollen from the anthers and therefore are not efficient pollinators of the plant.

This species is most abundant in the western Mojave Desert and becomes less abundant in southern cismontane California.

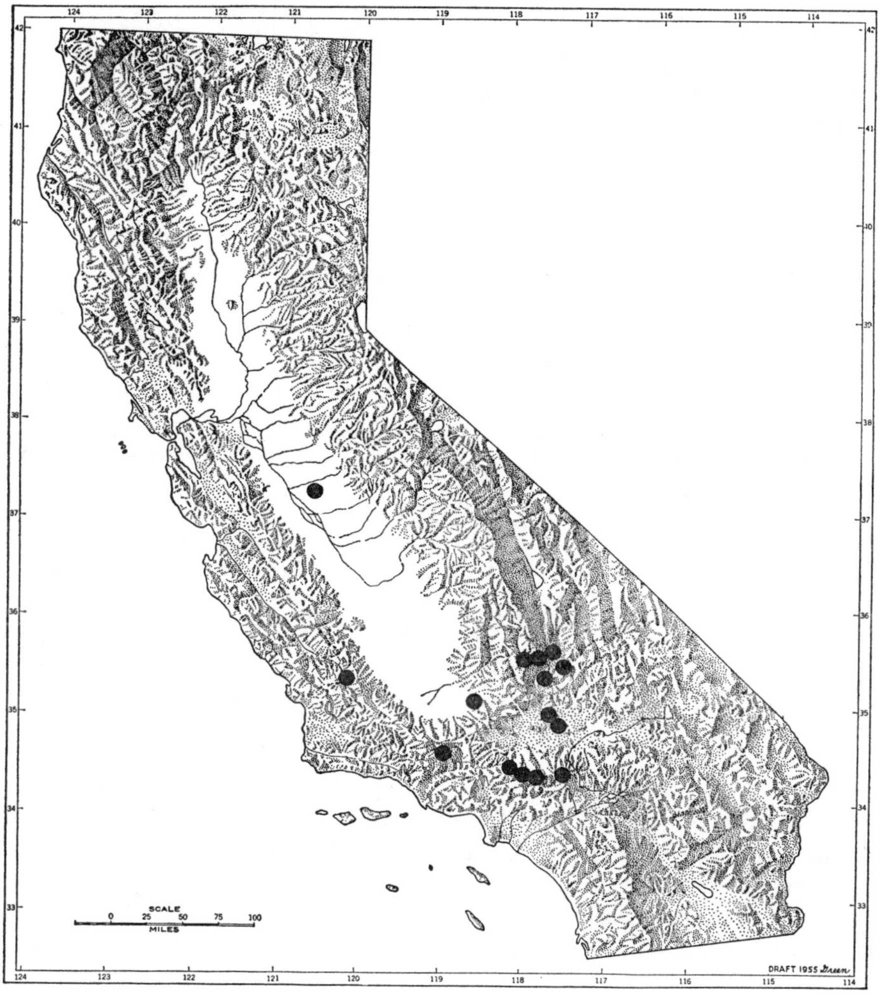

Fig. 8. Distribution of *Andrena* (*Diandrena*) *sperryi.*

Andrena (*Diandrena*) *anatolis* Linsley & MacSwain

Andrena anatolis, one of the two members of the Chalybioides Group (Thorp, 1969) associated with *Camissonia,* occurs from northwestern Baja California, Mexico, north to Ventura and Kern counties, California (fig. 9). Its principal pollen source is *Camissonia bistorta,* but it also takes pollen from *C. campestris* and *C. cheiranthifolia suffruticosa.* Our observations are based on 86 females. We have female records from late March to early May. Thorp (1969) records the females as early as early March and the males from mid-February to late April.

Nectar records for the males cited by Thorp (1969) include: *Camissonia bistorta, Ceanothus, Calandrinia, Cryptantha,* and *Potentilla.* Most of these records were with *Calandrinia* and *Cryptantha.*

The females collected pollen principally from *Camissonia bistorta* and also from

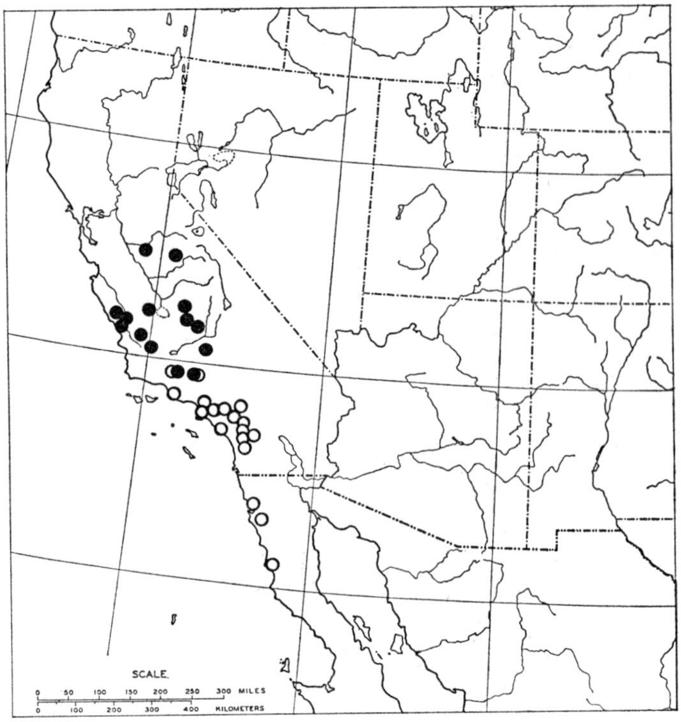

Fig. 9. Distribution of *Andrena* (*Diandrena*) *eothina,* dots; and *Andrena* (*D.*) *anatolis,* circles.

C. cheiranthifolia suffruticosa and *C. campestris.* Despite their small size, the few time records available indicate that the pollen foraging period is from early morning to midmorning.

The two largest population samples available are from near Sage, Riverside County, and Newport Beach, Orange County. Although the latter do not have flower labels, they were probably taken in association with *C. cheiranthifolia.*

The southernmost record for this species is near San Quintín, Baja California. This is about 80 miles south of the limit of distribution for *Camissonia bistorta.* At this locality a female without pollen was taken on *C. crassifolia,* a plant species which continues much farther south. If the bee can utilize the pollen of this plant, it may also occur farther southward.

Andrena (*Diandrena*) *eothina* Linsley & MacSwain

Andrena eothina, the closest relative of *A. anatolis,* is known only from Merced County to Ventura County, California (fig. 9). It collects pollen principally from *Camissonia campestris* and *C. sierrae sierrae,* and also from *C. contorta* and *C. graciliflora.* Our observations are based on 507 females and 49 males. Both sexes are on the wing from early March to mid-May.

Males of *A. eothina* were rarely taken in association with the flowers of *Camissonia.* Only four males were encountered on *C. campestris* and one on *C. sierrae.* Thorp (1969) records 14 additional males taking nectar from flowers of the genera

Tropidocarpum, Agoseris, Calandrinia, and *Amsinckia* with most of the captures made on the latter two genera.

Burrows of *A. eothina* are constructed in light sandy soils. They are in sparsely vegetated areas within the range of the pollen sources of the females. The entrance is surrounded by a concentric tumulus and is in bare flat ground or associated with the bases of *Erodium* plants. The plaster cast of a nest excavated near Bakersfield revealed that the tunnel from the entrance descended at a slight angle from the vertical and contained an enlarged area about halfway between the entrance and the bend to the lateral. The main shaft descended about 12.5 mm to the bend and the lateral shaft was about 11.5 mm long. The diameter of the main shaft varied from 3 mm to 3 by 5 mm; the diameter of the lateral shaft was about 5 mm. The enlarged area was about 7 by 9 mm in diameter and 9 mm long. An incomplete nest taken near Oakhurst also had a similar enlarged area, but it was closer to the entrance. These enlarged areas were not associated with irregularities in the soil and thus may not be artifacts. If the enlarged area is an integral part of the nest, its function is not known. The cells were found at a depth of 19 cm and they were thinly lined with a layer of waterproof material.

Females collect pollen primarily from *Camissonia campestris* and *C. sierrae,* foraging from early to midmorning. It usually appears on the flowers after *A. foxii* and *A. pasta,* but before *A. cyanosoma* and *A. macswaini* (tables 1, 3, 4, 5, 6, 7). Females visit nearly every flower, apparently scavenging from those visited by earlier bees or finding unvisited flowers. They move in a circular path around the flower visiting the stamens in sequence, but usually not contacting the erect pistil. Thus they do not normally pollinate the flowers.

The shape of the pollen ball and the position of the egg in the cells of *A. eothina* are essentially as those described above for *A. chalybaea* and *A. foxii.* The pollen ball is about 3.5 mm in greatest diameter and 2.5 mm in height.

This bee is most abundant in the southern end of the Central Valley and the valleys of the adjacent mountain ranges. It appears to be dominant only in the vicinity of Oakhurst in association with *C. sierrae.*

Clinal variation in color of integument and pubescence has been discussed for this species by Thorp (1969). Females of southern populations are uniformly dark blue with blackish pubescence while populations to the north contain increasing numbers of lighter, more greenish females with paler pubescence.

Andrena (Diandrena) cyanosoma (Cockerell)

Andrena cyanosoma appears most closely related to *A. macswaini,* although placed in a species group by itself by Thorp (1969). It occurs from northwestern Baja California, Mexico, north to Madera County, California (fig. 10). Females collect pollen principally from *Camissonia campestris* and have been found also in association with *C. sierrae, C. bistorta,* and *C. cheiranthifolia suffruticosa.* Thorp (1969) also records a female on *C. pallida.* Our observations are based on 102 females and 9 males. Both males and females are primarily found from early March to mid-May, although Thorp (1969) records the females flying as late as mid-June.

Eight of the males taken in this study were associated with *Camissonia campes-*

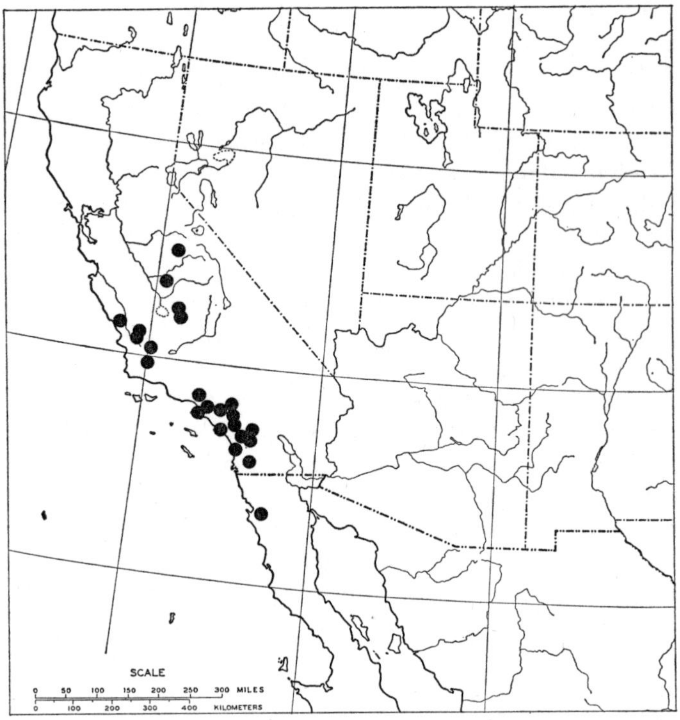

Fig. 10. Distribution of *Andrena* (*Diandrena*) *cyanosoma.*

tris and *C. sierrae.* Thorp (1969) also records 12 males taken at *Cryptantha* flowers and one at the flowers of *Rhus.*

Females collect pollen exclusively from flowers of *Camissonia.* They forage from early morning to midmorning and usually are among the latest of the *Camissonia* oligoleges to appear at the flowers (tables 5, 7).

This species appears to be most abundant in the southern end of the Central Valley, but is not dominant nor very abundant at any locality in relation to other oligoleges of *Camissonia.*

Andrena (*Diandrena*) *macswaini* Linsley

Andrena macswaini appears most closely related to *A. cyanosoma* (Linsley, 1960), although placed by Thorp (1969) in a separate species group. It is known only from a few localities from Kern County to Madera County, California (fig. 11). It collects pollen only from *Camissonia campestris* and the closely related *C. sierrae.* Our observations are based on 355 females and 20 males. Females were taken from late February to mid-May, but males were encountered only in early March.

Males were rarely found visiting flowers. Thirteen were taken on *Camissonia campestris* and two on a species of Compositae. Males were most commonly encountered cruising the aggregated nests of the females. One male was observed

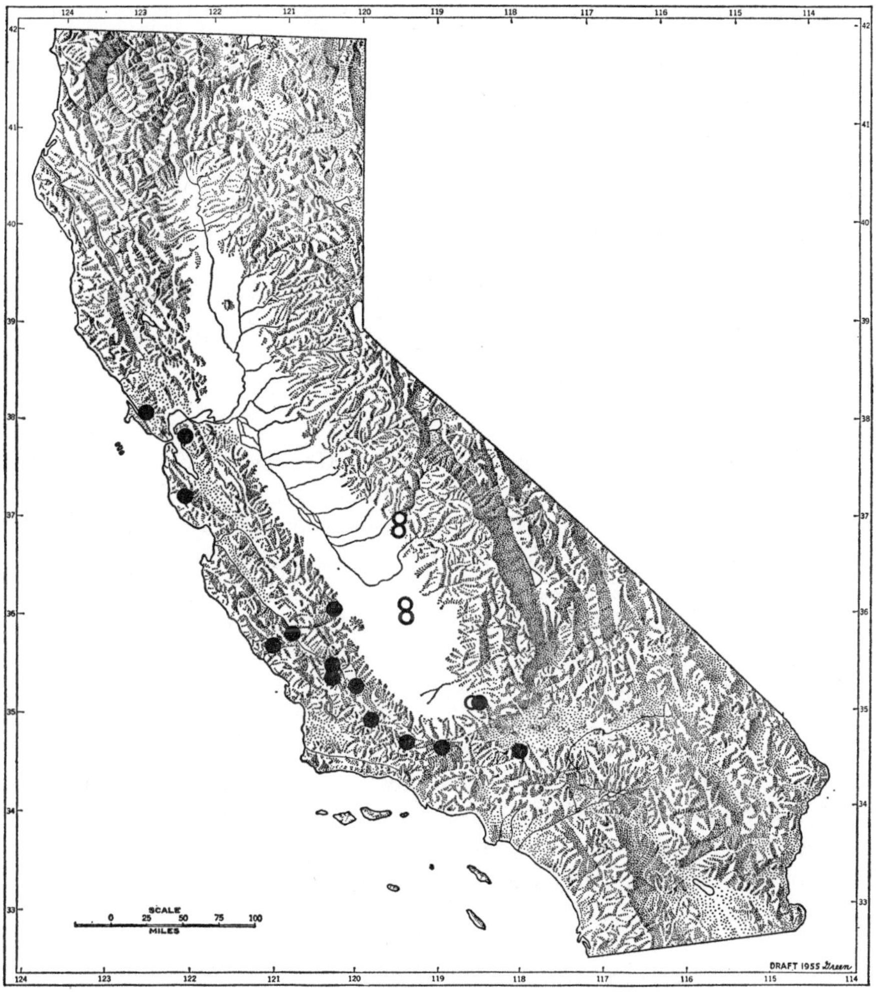

Fig. 11. Distribution of *Andrena* (*Diandrena*) *apasta*, dots; and *Andrena* (*D.*) *macswaini*, circles.

crawling under a stone in the middle of the afternoon when most male activity had ceased. Presumably this is where it spent the night.

Burrows of *A. macswaini* are constructed in deep sandy soils. They differ from the general pattern of other species of *Diandrena* previously discussed in both microhabitat and in depth. They occur on flat ground in aggregations associated with distinct depressions (pl. 2). They are most commonly found in small gullies resulting from water runoff. In one instance the serial depressions formed by a tracked vehicle were utilized for burrow initiation. The shape of the large tumulus surrounding the entrance of the nest is dictated by the size and shape of the depression in which it is constructed, but it is often concentric. The burrows tended to spiral downward from the entrances for the first 7 to 10 mm and then continued vertically. As indicated by their large tumuli the nests were very deep. The cells of most other *Diandrena* range from 10 to 35 cm below the surface, while in *A. macswaini* a plaster cast was excavated to a depth of 78 cm without reaching the level of the presumed lateral shaft or cells (pl. 2).

Females collect pollen only from *Camissonia campestris* and *C. sierrae*. The daily foraging period usually begins after those of most of the other oligoleges (tables 1, 3, 7). One female obtained a full load of pollen from four flowers in 90 seconds. Females move in a circular path around the flower, moving from stamen to stamen and frequently contacting the stigma.

This species occurs only in a very restricted area of southcentral California. The largest single sample of 79 females was taken on 1 April 1962 at 18 miles east of Bakersfield.

Andrena (*Diandrena*) *apasta* Linsley & MacSwain

Andrena apasta is not closely related to other *Camissonia* oligoleges. It occurs from Los Angeles County north to Marin County (fig. 1). It collects pollen principally from *Camissonia campestris* and has occasionally been found associated with *C. ovata* and one female was recorded taking nectar from *C. pallida* in the Mojave Desert (Linsley, MacSwain, and Raven, 1964). Our observations are based on 227 females and 31 males. Males commence flying in late February and females in early March, and both are on the wing until early May.

Males are rarely encountered on the flowers of *Camissonia*. Most were taken while while cruising in the vicinity of these plants. Thorp (1969) records 5 males collected on flowers of *Monolopia, Agoseris,* and *Descurainia.*

Females collect pollen exclusively from flowers of *Camissonia,* principally from *C. campestris.* They forage from early morning to midmorning. In Monterey and San Luis Obispo counties where the bee is most abundant, it is one of the earliest visitors to these flowers, and is a dominant *Camissonia* oligolege in this area (tables 4, 6).

Andrena (*Onagrandrena*)

Nine species of *Andrena* (*Onagrandrena*)—*A. blaisdelli* Timberlake, *A. camissoniae* Linsley & MacSwain, *A. convallaria* Linsley & MacSwain, *A. deserticola* Timberlake, *A. eulobi* Linsley & MacSwain, *A. furva* Linsley & MacSwain, *A. oenotherae* Timberlake, *A. oraria* Linsley & MacSwain, and *A. vespertina* Linsley & MacSwain—occur in cismontane California and northwestern Baja California, Mexico. The first seven collect pollen principally from yellow-flowered, morning-opening species of *Camissonia* while the last-named species collects pollen from the white-flowered, evening-opening *Camissonia boothii* subsp. *decorticans.* Two of these bees, *A. deserticola* and *A. furva,* also occur on the Mojave Desert and have been discussed by Linsley, MacSwain, and Raven (1964). *Andrena deserticola* occurs more abundantly along the western edge of the Mojave Desert and will not be treated in detail here. The principal field observations on these species were conducted between 1959 and 1964. The species of the Oenotherae species complex were treated by Linsley and MacSwain (1963).

Because of the superficial similarity of these bees, keys are offered to help distinguish males and females of *Onagrandrena* from cismontane California and Baja California. Males of *A. camissoniae* and *A. furva* are not known.

KEY TO ONAGRANDRENA OLIGOLECTIC ON CAMISSONIA IN CISMONTANE CALIFORNIA AND BAJA CALIFORNIA

FEMALES

1 Pubescence black 2
Pubescence pale ochraceous; length of body 11 mm, anterior wing 9 mm .... *camissoniae*
2 (1) Propodeal enclosure coarsely rugose; most mesoscutal punctures less than one diameter apart 3
Propodeal enclosure rugulose; most mesoscutal punctures at least one diameter apart; length of body 10.5–12 mm, anterior wing 8–9 mm .... *deserticola*
3 (2) Metasomal terga with a very broad impunctate apical margin, covering more than one-half of apical impression medially. 4
Metasomal terga with a narrow impunctate apical margin, covering less than one-half of apical impressions 5
4 (3) Mesosoma clothed with short, dense, brown, heavily plumose hairs, with a few longer black hairs intermixed; length of body 10.5–11 mm, anterior wing 7 mm .... *furva*
Mesosoma clothed with long, erect, black hairs; length of body 9.5–11 mm, anterior wing 7–8 mm .... *blaisdelli*
5 (3) Clypeus with a distinct although frequently feeble and incomplete median longitudinal smooth line 6
Clypeus without a median longitudinal smooth line 7
6 (5) Wings almost clear; labrum with apical process short, one and one-half times as broad as long; length of body 10.5–13 mm, anterior wing 8–9.25 mm .... *eulobi*
Wings tinted with blackish; labrum with apical process elongate, one and one-half times to twice as long as broad; length of body 11–13.5 mm, anterior wing 8–9 mm .... *oraria*
7 (5) Mesoscutum subopaque, reticulations between punctures well developed, enclosed areas more or less round, elevated 8
Mesoscutum feebly shining, reticulations between punctures weakly to well developed, enclosed areas elongate, narrowly impressed; length of body 11.5–13 mm, anterior wing 8.5–9.25 mm .... *vespertina*
8 (7) Labrum with process more than one and one-half times as long as wide, parallel-sided; length of body 9.5–12 mm, anterior wing 7.5–8.5 mm .... *oenotherae*
Labrum with process as long or one and one-half times as long as wide, sides diverging from base to apex; length of body 11.5–13.5 mm, anterior wing 8.25–9 mm .... *convallaria*

MALES

1 Pubescence of metasoma pale 2
Pubescence of metasoma predominantly black 4
2 (1) Pubescence of face all pale 3
Pubescence of face white centrally, black laterally; length of body 9.5–10 mm, anterior wing 7.5–8 mm .... *eulobi*
3 (2) Posterior impression of metasomal tergites impunctate on apical one-half; length of body 10 mm, anterior wing 7.5 mm .... *blaisdelli*
Posterior impression of metasomal tergites impunctate on apical one-third; length of body 10 mm, anterior wing 7 mm .... *deserticola*
4 (1) Facial pubescence usually black, rarely with some pale hairs on apex of clypeus, on genae and on antennal scape .... *oenotherae, convallaria,* and *oraria*
Facial pubescence predominantly black, pale hairs on apex of clypeus and on genae behind mandibles; length of body 10.5 mm, anterior wing 8 mm. .... *vespertina*

Andrena (*Onagrandrena*) *blaisdelli* Cockerell

Andrena blaisdelli is known only in southern California from southwestern Riverside County north to San Luis Obispo County and northern Kern County (fig. 12). Cockerell (1924) named this species from a female from San Diego. Its

Fig. 12. Distribution of *Andrena (Onagrandrena) blaisdelli*, dots; and *Andrena (O.) camissoniae*, circle.

principal pollen source is *Camissonia campestris*, but it has also been found taking pollen from *C. bistorta*, *C. californica*, and *C. contorta*. Our records are based on 100 females and 2 males. Females are on the wing from late March to mid-May. The two males were taken in early May, on flowers of *Cryptantha intermedia*.

Females collect pollen from *Camissonia*, principally *C. campestris*, in the early morning (tables 6, 8). Of 79 females taken on flowers of this species, 66 were gathering pollen. On other species of *Camissonia*, ten of 15 females collected from *C. bistorta* were gathering pollen; the only female from *C. californica* was gathering pollen as was one of the two taken on *C. contorta*. Females presumably meet most of their nectar requirements from *Camissonia* flowers, but occasionally they have been found taking nectar from other flowers. One nectar-collecting female was taken from each of the following plants: *Eriophyllum* sp., *E. confertiflorum,* and *Cryptantha intermedia.*

In addition to the records listed by Linsley and MacSwain (1955), we can now add the following. Kern Co.: Woody, one female (N)[5] on 24 April 1949. Los Angeles Co.: Claremont, two females (P),[5] on *Camissonia bistorta* at 8:27 and 10:29 on 17 April 1959; La Mirada, one female (N) on 21 March 1940. Riverside Co.: Temecula, 7 miles east, 11 females (8P) on *C. bistorta* between 6:29 and 7:29 on 23 April 1959. San Luis Obispo Co.: Santa Margarita, 5 miles northeast, four females (4P) on *C. campestris* between 6:40 and 7:10 between 2 and 4 May 1964; Simmler, 10 miles west, 53 females (42P) on *C. campestris* between 6:00 and 9:10 between 5 and 13 May 1962, one female (N) on *Eriophyllum* sp. at 8:30 on 6 May 1962. Ventura Co.: Hungry Valley, 5 miles south of Gorman, 22 females (20P) on *C. campestris* between 5:55 and 7:58 on 7 May 1959 and 10 April 1960, one female (N) on *Eriophyllum confertiflorum* on 6 May 1959.

Andrena (*Onagrandrena*) *convallaria convallaria* Linsley & MacSwain

Andrena camissoniae is known only from the type series of four females taken 28 miles northwest of New Cuyama, Santa Barbara County, California (fig. 12) on 8 June 1963 between 6:50 and 7:30 on *Camissonia campestris.* This species, distinctive in its ochraceous pubescence, the short, blunt labral process and the sculpturing of the propodeal enclosure, has tentatively been assigned to the *A.* (*O*) *oenotherae* complex by Linsley and MacSwain (1968).

Andrena (*Onagrandrena*) *convallaria convallaria* Linsley & MacSwain

Andrena convallaria as known, comprises two subspecies (Linsley and MacSwain, 1963). *Andrena c. subhyalina* Linsley and MacSwain is known presently only from the western edge of the Mojave Desert (Linsley, MacSwain, and Raven, 1964). The nominate subspecies occurs in cismontane California between Antioch and Bakersfield in the Central Valley and between San Luis Obispo County and Ventura County in the inner Coast Ranges (fig. 13), where the principal pollen source is *Camissonia campestris* although it has also been taken occasionally with pollen of *C. cheiranthifolia, C. californica,* and *C. contorta.* Our observations involve 589 females and 143 males. Both males and females are active from early March through late May with some females extending their activity into early June.

Males were commonly encountered taking nectar from flowers of *Camissonia* (100 were taken on *C. campestris*). Other floral records include: 26 on *Isomeris arborea,* three on *Malacothrix,* one on *Encelia actonii,* and one on *Lasthenia.*

Females collect pollen and nectar from *Camissonia.* Our records include: 377 with pollen and 129 nectar on *C. campestris;* two pollen on *C. californica;* three pollen and five nectar on *C. contorta;* one pollen, three nectar on *C. c. cheiranthifolia;* and two nectar on *C. micrantha.* Nonpollen records on other plants include: one each *Lupinus, Brassica,* and *Isomeris arborea.* Females collect pollen from early morning to midmorning (tables 1, 3–8). They orient across the center of the flower contacting the stigma and stamens when gathering pollen and do not

[5] P = with pollen; N = without pollen.

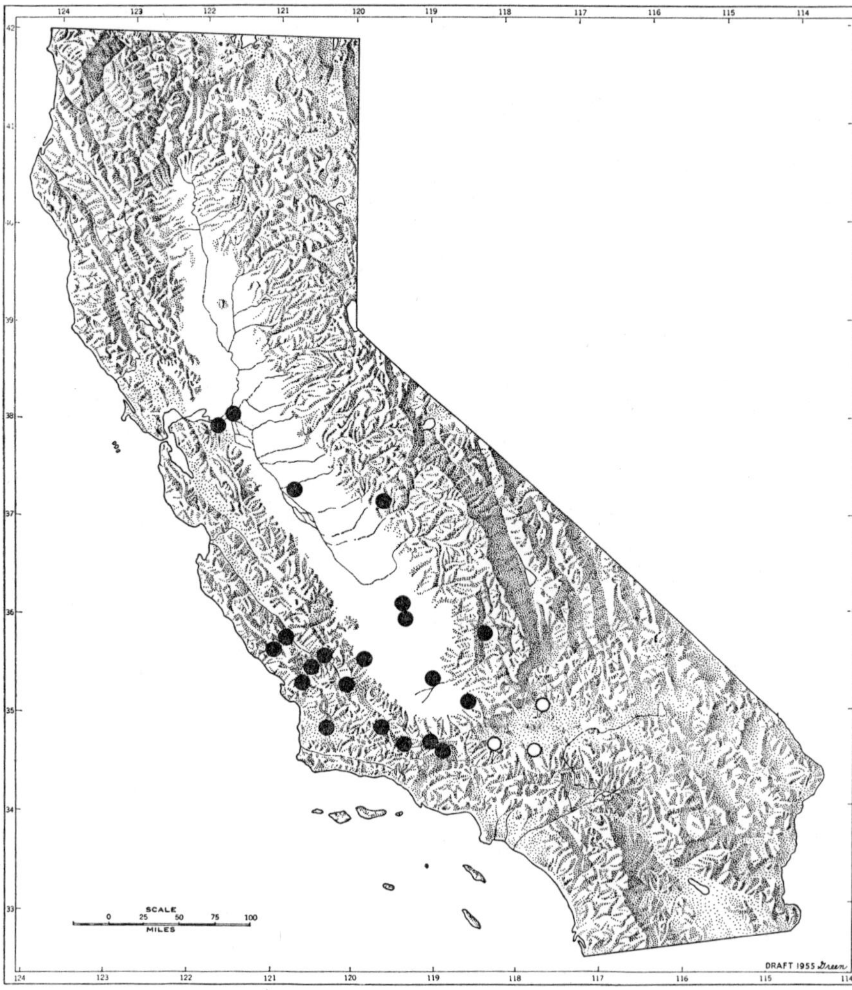

Fig. 13. Distribution of *Andrena* (*Onagrandrena*) *convallaria:* subspecies *convallaria,* dots, subspecies *subhyalina,* circles.

move in a circular path around the flower. In taking nectar they alight on the petals on the lower side of the flower and push their mouthparts into the centrally located nectary.

In addition to the records published by Linsley and MacSwain (1963) we have the following California records. Kern Co.: Bakersfield, 7.3 miles north, six females (P) on *C. campestris* between 7:10 and 7:40 on 3 April 1962. Madera Co.: Oakhurst (cemetery) six males, 19 females (12P) on *C. sierrae* between 7:20 and 9:15 on 15 and 16 May 1963; Oakhurst, 3.7 miles west, 19 females (17P) on *C. sierrae* between 7:23 and 9:20 on 15 May 1963. Monterey Co.: Bradley, 1.5 miles west, four females (P) on *C. campestris* between 8:00 and 9:50 on 3 May 1963. Sacramento Co.: Brannan Island State Park, four females (1P) on *C. c. cheiran-*

thifolia between 6:45 and 7:45, and one female (N) on *Brassica* on 20 May 1965. San Luis Obispo Co.: Bee Rock, 1.8 miles west, one male, six females (4P) on *C. campestris obispoensis* between 6:45 and 9:30 on 6 May 1963; Santa Margarita, 4.5 miles northeast two females (N) on *C. micrantha* between 8:03 and 10:55 on 28 April 1959 and 2 May 1962, four females (3P) on *C. contorta* between 6:54 and 10:58 on 22 May 1959, and 2 and 4 May 1962, two females (P) on *C. californica* between 9:32 and 9:34 on 28 April 1959, one female (N) (no host) at 10:48 on 28 April 1959; Santa Margarita, 5 miles northeast, one female (N) on *C. campestris* at 11:35 on 5 May 1962. Santa Barbara Co.: Cuyama, two males on 19 March 1958; New Cuyama, two males on 28 March 1957; 28 miles northwest of New Cuyama, 18 females (14P) on *C. campestris* on 8 June 1963 between 6:50 and 7:40. Tulare Co.: Fairview, 1 mile south, five females (4P) on *C. campestris* between 9:00 and 9:15 on 30 April 1964; Tipton, 2 miles south, two females (P) on *C. campestris* between 8:00 and 8:30 on 3 April 1962; Tulare Airpark, three females (two with pollen) on *C. campestris* between 8:25 and 8:55 on 6 April 1964, one male, 63 females (58 with pollen) on *C. campestris* between 6:17 and 7:57 on 16 April 1964; Tulare, 4.5 miles south, three females taking pollen on *C. campestris* between 8:52 and 9:04 on 29 March 1960. Ventura Co.: Quatal Canyon, northwest corner of County, one male on *Encelia actonii* at 12:30 on 9 May 1959.

Andrena (*Onagrandrena*) *deserticola* Timberlake

Andrena deserticola occurs principally on the western edge of the Mojave Desert, but has been taken rarely in cismontane California in the extreme southeastern part of the San Joaquin Valley and northeastern Ventura County (fig. 14). The species has been discussed by Linsley, MacSwain, and Raven (1964). *Camissonia campestris* is the only pollen source we have found for *A. deserticola* in cismontane California. Our records are based on six females taken between early March and early May.

Females collect pollen in the early morning (table 1). Four of the six females were taken gathering pollen and one collecting nectar from *C. campestris;* the remaining female was collecting nectar from *Isomeris arborea.* One of the pollen-gathering females from Hungry Valley was taken at 6:38.

In addition to the records given by Timberlake (1937) and Linsley and MacSwain (1955) we have the following California records. Kern Co.: Bakersfield, 18 miles east, one female without pollen on 9 March 1959, three females with pollen on *C. campestris* between 6:38 and 8:00 on 9 April 1960 and 1 and 2 April 1962. Ventura Co.: Hungry Valley, 5 miles south of Gorman, one female taking pollen from *C. campestris* at 6:38 on 7 May 1959.

Andrena (*Onagrandrena*) *eulobi* Linsley & MacSwain

Andrena eulobi is known only from Baja California, Mexico (fig. 14). The type series of 37 females is from 14 miles south of San Quintín on 30 March 1962 at the flowers of *Camissonia crassifolia.* Only 4 of the 37 females lack pollen from this plant. Another series from Guerrero Negro, just at the line between the State (Norte) and the Territory (Sur) contains four males (one sleeping in flower)

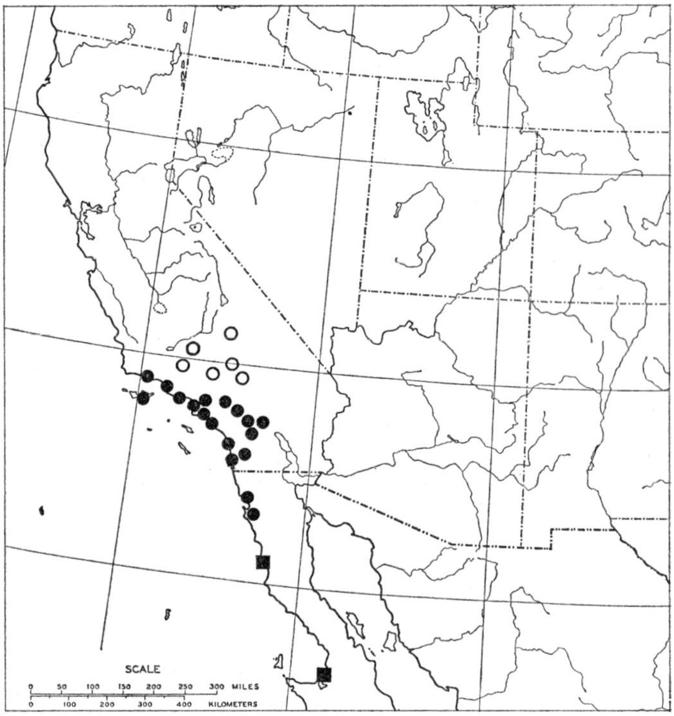

Fig. 14. Distribution of *Andrena* (*Onagrandrena*) *deserticola,* circles; *Andrena* (*O.*) *oenotherae,* dots; and *Andrena* (*O.*) *eulobi,* squares.

and 13 females (two with small loads of pollen) from *C. crassifolia* on 29 March 1972 between 6:10 and 7:00. Linsley and MacSwain (1963) assigned this species to the *A.* (*O.*) *oenotherae* complex.

Andrena (*Onagrandrena*) *furva* Linsley & MacSwain

Andrena furva is known only from San Diego County north to San Luis Obispo County (fig. 15). Its principal pollen source appears to be *Camissonia campestris,* although it has been observed gathering pollen from *Clarkia cylindrica.* Our records are based on 20 females taken between 17 April and 13 June; the male is unknown.

This species does not appear to be well adapted as an oligolege of *Camissonia* because it flies late in the season and late in the day (most of our records are between 10:30 and 11:30). Linsley and MacSwain (1961) suggested that *A. furva* might be closed to *A. blaisdelli.* However, Thorp (1969: 43) and MacSwain, Raven, and Thorp (1973) relate *A.furva* to *A.omninigra* Viereck, an oligolege on *Clarkia.* They point out the significance of this relationship in light of the fact that *A. furva* has been taken on *Clarkia cylindrica* with pure loads of *Clarkia* pollen or with mixed loads of *Clarkia* and *Camissonia* pollen. Six females taken at Frey Creek, Pauma Valley, San Diego County, on 17 and 19 April 1962, were without any flower association, but all six had *Camissonia* pollen in their scopae.

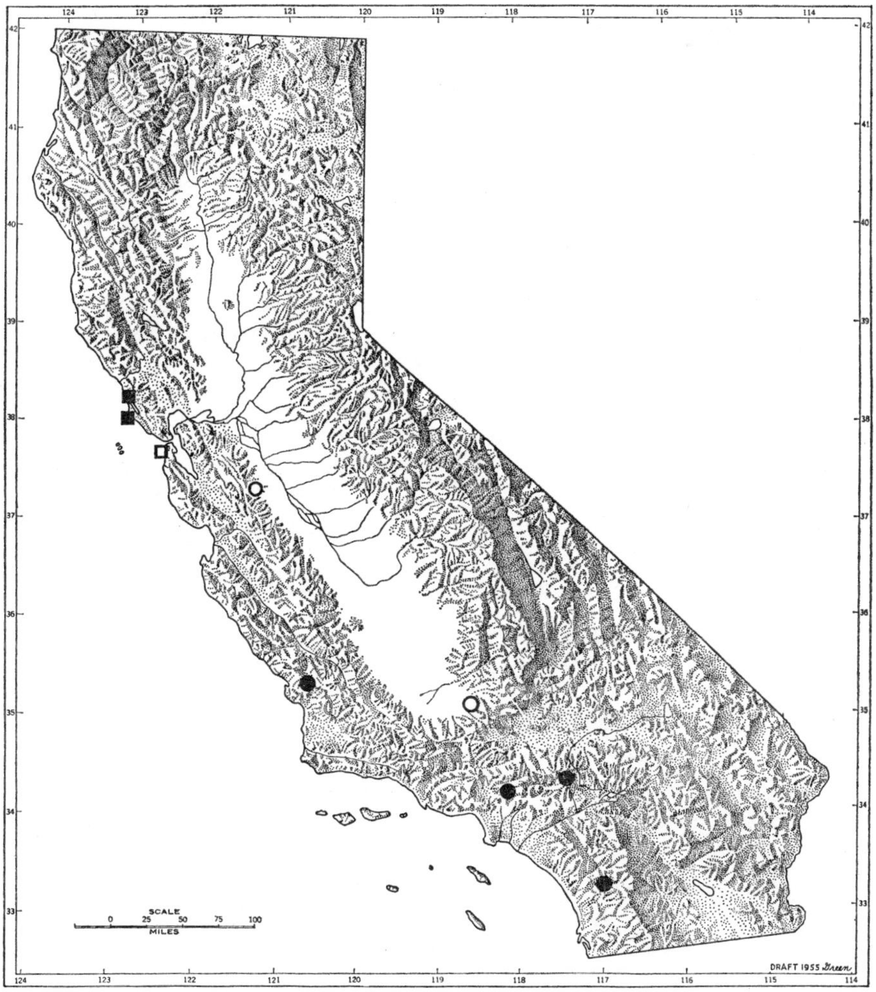

Fig. 15. Distribution of *Andrena (Onagrandrena) oraria:* subspecies *oraria,* solid squares, subspecies *actitis,* hollow square; *Andrena (O.) vespertina,* circles; *Andrena (O.) furva,* dots.

In addition to the records given by Linsley and MacSwain (1961) and Linsley, MacSwain, and Raven (1964) the following subsequent California records are biologically important. San Luis Obispo Co.: Santa Margarita, 5 miles northeast, five females (4P) on *C. campestris* between 10:30 and 12:05 on 4 and 5 May 1962, one female (P mixed with *Clarkia*) on *C. campestris* 13 June 1963, four females (3P *Clarkia*, 1P *Clarkia* mixed with *Camissonia*) on *Clarkia cylindrica* between 10:00 and 12:00 on 13 June 1963.

Andrena (Onagrandrena) oenotherae Timberlake

Andrena oenotherae occurs on Santa Cruz Island, along the mainland coast, and into the interior valleys from Santa Barbara County south to Santo Tomás, Baja California, Mexico (fig. 14). Its principal pollen source is *Camissonia cheiranthifolia cheiranthifolia* on Santa Cruz Island, and *C. c. suffruticosa* along the mainland coast while the more interior populations collect from *C. bistorta.* Our records are based on 574 females and 145 males. Females are on the wing from 13 March to 9 July, while the males have been taken from 13 March to 25 June.

A majority of the males (117) have been taken at the flowers of *Camissonia cheiranthifolia suffruticosa* indicating that the flowers may serve as the site for mating as well as the principal source of pollen for the females. Two males were collected at the flowers of *Cakile edentula.*

This species appears to be most abundant along the coastal strand. Here 349 pollen-gathering females and 118 nectar-collecting females were collected at flowers of *C. cheiranthifolia suffruticosa.* Among the inland population 24 pollen-gathering females and three nectar-collecting females were taken at flowers of *C. bistorta.* In addition 15 nectar-collecting females have been taken at the flowers of *Chaenactis* sp. with two of these having pollen from *Camissonia* in their scopae. These bees forage for pollen early in the morning, most of our records are between 6:30 to 8:30 (table 9).

Biological information attributed to this species by Linsley, MacSwain, and Smith (1955) and Linsley and MacSwain (1956) actually pertains to *A* (*O.*) *boronensis* Linsley and MacSwain (Linsley, MacSwain, and Raven, 1964). In their discussion of the "oenotherae complex" Linsley and MacSwain (1963) describe several new related taxa, further restrict the definition of *A. oenotherae,* list the localities from previous authors (Timberlake, 1937; Linsley, 1938; Linsley and MacSwain, 1955), and new localities which are assignable to the present species concept. Additional records accumulated since that time include the following. Mexico. Baja California del Norte: Descanso Bay, one male, 23 March 1941 (Cockerell, 1941); Ensenada, 3 miles south, 18 females (14P) on *C. cheiranthifolia suffruticosa* between 7:06 and 8:53 on 29 March 1962; Santo Tomás, 7 miles north, five females (P) on *C. bistorta* between 7:20 and 9:25 on 31 March 1962. California. Riverside Co.: Homeland, two females (P) on *C. bistorta* at 9:06 and 9:07 on 14 March 1962; Sage, 5 miles south, 13 females (12P) on *C. bistorta* between 6:30 and 7:48 on 17 April 1965. San Bernardino Co.: Alta Loma, three females (1P) on *C. bistorta* between 7:08 and 7:35 on 26 March 1962. San Diego Co.: Carlsbad, 2 miles south, 18 males, 39 females (31P) on *C. cheiranthifolia suffruticosa* mostly before 9:00 on 2 April 1961; Mission Bay, one male, *Melilotus alba* between 8:55 and 9:15 on 25 June 1963, two males, eight females (6P) on *C. cheiranthifolia suffruticosa* between 7:15 and 8:55 on 25 June 1963, two females (P) on *C. cheiranthifolia suffruticosa* on 9 July 1963; Solana Beach, two males, two females (N) on *Cakile edentula* on 19 June 1963. Santa Barbara Co.: Christie Beach, Santa Cruz Island, one female 5 May 1968, two females 27 April 1969.

Andrena (*Onagrandrena*) *oraria* Linsley & MacSwain

Andrena oraria consists of two subspecies which are restricted to the coastal strand from San Francisco County to Sonoma County (fig. 15). A key for the separation of the females of the two subspecies is presented in Linsley and MacSwain (1963). Both collect pollen principally from *Camissonia cheiranthifolia cheiranthifolia.*

Andrena (*Onagrandrena*) *oraria oraria* Linsley & MacSwain

Andrena o. oraria has been taken at two localities north of the Golden Gate entrance to San Francisco Bay—at the type locality, 1 mile northeast of Pt. Reyes Lighthouse, Marin County, and at Dillon Beach, Sonoma County. It takes pollen from both *C. c. cheiranthifolia* and *C. ovata.* Our records are based on 19 females and 20 males. Females are on the wing from 11 April (erroneously recorded as 4 February in Linsley and MacSwain, 1963) to 18 May, while males are on the wing from 14 April to 18 May.

One male was excavated from its cell on 10 October 1959, and two females were excavated from their cells on 12 February 1939. Thus, this species appears to overwinter in the adult stage as is typical of most vernal *Andrena* (see Thorp, 1969: 18). Most of the males (14) were taken at nest sites indicating that this is the site for mating.

Two pollen-gathering females were taken at flowers of *C. c. cheiranthifolia* and two pollen-gathering females and one nectar-collecting female were taken at flowers of *C. ovata.*

In addition to the records listed by Linsley and MacSwain (1963) we have a record of three males and one female taking nectar from *C. c. cheiranthifolia* on 18 May 1963 at Dillon Beach, Sonoma County.

Andrena (*Onagrandrena*) *oraria actitis* Linsley & MacSwain

Andrena o. actitis is known only from two localities south of the Golden Gate entrance to San Francisco Bay in San Francisco County—Bakers Beach State Park, and the southwest corner of Fleishhacker Zoo (Golden Gate Park). It collects pollen only from *Camissonia c. cheiranthifolia.* Our records are derived from 38 females and 14 males taken on 19 and 21 May 1959.

The male from Bakers Beach was taken at a nest site at 11:15 on 19 May.

Of the 38 females taken at flowers of *C. c. cheiranthifolia,* 25 were gathering pollen.

Andrena (*Onagrandrena*) *vespertina* Linsley and MacSwain

Andrena vespertina is known only from two widely separated areas—18 and 20 miles east of Bakersfield, Kern County; and Del Puerto Canyon, 21 miles west of Patterson, Stanislaus County (fig. 15). The females collect pollen only from the white-flowered, evening-opening *Camissonia boothii decorticans.* Our observations involve 66 females and 99 males. The females fly from 9 March to 12 June; males are on the wing from 9 March to 21 May.

Males were most frequently found on or cruising flowers of *Isomeris arborea* in the morning. Of the 62 males taken at flowers, 46 were taken in association with this plant. One male was taken on the flower of an unidentified composite. The remaining 15 were taken at the flowers of *Camissonia boothii decorticans* within a few hours of sunset.

Females collect pollen exclusively from *C. b. decorticans* and during the two hours preceding sunset (table 10). The plants grow principally on slopes and are usually positioned so that they are in the shade for the entire pollen-collecting period. Of the 38 females associated with flowers, 16 nectar-collectors and 17 pollen-gatherers were taken from *C. b. decorticans,* and five nectar-collectors were taken at *Isomeris arborea.*

We have observed two other populations of *C. b. decorticans* in our area—10 miles west of Simmler, San Luis Obispo County; and 2 miles north of Paicines, San Benito County—without encountering *A. vespertina.* Perhaps these plant populations are principally moth pollinated as appears to be the case for desert populations (Linsley, MacSwain, and Raven, 1964: 82).

Among the paratype series designated by Linsley and MacSwain (1961) all 19 females collected in the morning on "*Oenothera dentata parishii*" should be transferred to *A. c. convallaria.* Additional records from the type locality include: 26 males cruising *Isomeris arborea* between 10:30 and 15:40 on 31 March, 1 and 2 April 1962; and 12 males, two females (N trying to force open flowers) on *C. boothii* between 16:45 and 17:02 on 31 March 1962.

In addition to the above records we have: one male, 11 females (6P) on *C. boothii* between 16:45 and 20:41 on 20 and 21 May and 12 June 1967; and one female (N) seen on *C. boothii* at 6:31 on 21 May 1967 at Del Puerto Canyon, 21 miles west of Patterson, Stanislaus County.

OTHER BEE VISITORS OF CAMISSONIA AND OENOTHERA

Hylaeus species

Two females of *Hylaeus cressoni pasadenae* (Cockerell) were taken at Prairies Fork, San Bernardino Mountains, San Bernardino County, on *Oenothera hookeri* in the morning of 1 August 1961. Additional undetermined specimens of *Hylaeus* were collected in association with the following plants: *Camissonia cheiranthifolia cheiranthifolia,* one female, at Grover City, San Luis Obispo County, on 28 April 1959 at 8:45; and *Oe. hookeri,* two females, at East Highland, San Bernardino County, on the morning of 27 July 1961.

Perdita soluta Timberlake

One male of *Perdita soluta* was collected at 6.8 miles southwest of Shandon, San Luis Obispo County, taking nectar from *Camissonia campestris* on 10 April 1961 between 8:50 and 9:00.

Halictus tripartitus Cockerell

Sixteen females of *Halictus tripartitus* have been collected gathering pollen from the following plants: *Camissonia campestris,* one at 2.5 miles south of Tipton, Tulare County, on 2 March 1961; *C. cheiranthifolia cheiranthifolia,* one at Grover City, San Luis Obispo County, on 28 April 1959 at 7:01; and *Oenothera hookeri,* one at Coldwater Canyon, San Gabriel Mountains, Los Angeles County, on 10 August 1961 at 12:00; two at Grass Valley, San Bernardino Mountains, San Bernardino County, on 20 July 1961 at 9:30, and ten at the same locality on 31 July 1961 between 8:30 and 10:00.

Lasioglossum species

Three subgenera and 16 species of *Lasioglossum* are represented in collections of visitors of *Camissonia* and *Oenothera* in cismontane California. This is in marked contrast to the three species each found in the Colorado and Mojave Deserts (Linsley, MacSwain, and Raven, 1963*a* and 1964). It is comparable, however, to the situation in the Great Basin (Linsley, MacSwain, and Raven 1963*b*) with which it shares seven species.

Lasioglossum (*Lasioglossum*) spp.

One female of *Lasioglossum* (*L.*) *sisymbrii* (Cockerell) was taken with pollen at the flowers of *Camissonia californica* in Mint Canyon, Los Angeles County, on 1 April 1959. This bee has also been taken at other species of Onagraceae in the Colorado Desert and in the Great Basin.

Seven females and two males of *L.* (*L.*) *pavonotum* (Cockerell) were found in coastal localities from Santa Barbara County to San Francisco County. All the females were collecting pollen from flowers of *Camissonia cheiranthifolia cheiranthifolia*. One male was also found on this plant while the other was taken on *C. c. suffruticosa* at Goleta, Santa Barbara County, on 2 May 1959. Females were collected from 28 April to 23 May 1959 and between 7:25 and 10:45 at the following localities: Grover City, San Luis Obispo County (3); Seaside, and 17 mile Drive, Monterey County (3); and Fleishhacker Zoo, southwest corner, San Francisco County (2).

Lasioglossum (*Evylaeus*) spp.

Seven species of the subgenus *Evylaeus* were found in association with flowers of *Camissonia* and *Oenothera* in cismontane California. Five of these were also found in similar relationships in the Great Basin (Linsley, MacSwain, and Raven, 1963*b*) and one of these five was also taken in the Colorado and Mojave Deserts (Linsley, MacSwain, and Raven, 1963*a* and 1964).

Three females of *L.* (*E*) *cooleyi* (Crawford) were collected with pollen from flowers of *Camissonia ovata* 5 miles north of Lakeport, Lake County, on 24 April 1963 between 10:00 and 11:00.

One female of *L.* (*E.*) *glabriventre* (Crawford) was found with a small load of pollen from flowers of *Camissonia s. sierrae* at a cemetery in Oakhurst, Madera County, on 16 May 1963 between 8:20 and 8:30. This bee has also been taken on *Clarkia* (MacSwain, Raven, and Thorp, 1973).

A total of five females and one male of *L.* (*E.*) *kincaidii* (Cockerell) were found on flowers of *Camissonia* as follows: one female without pollen, on *C. c. cheiranthifolia* at Grover City, San Luis Obispo County, on 28 April 1959 at 8:45; one female with pollen on *C. campestris* at 2 miles southwest of Livingston, Merced County, on 5 April 1959 at 9:10; two females, only one with pollen, on *C. contorta* at 2.5 miles south of Livingston on 20 and 26 March 1960 at 10:06 and 10:58; and one male and one female with a small load of pollen on *C. campestris* at the same site on 26 March and 2 April 1960 at 10:26 and 8:34.

Three females of *L.* (*E.*) *nigrescens* (Crawford) with pollen were found at flowers of *Camissonia ovata,* one at 5 miles north of Lakeport, Lake County, on 24 April 1963 between 10:00 and 11:00; and two others at 1.7 miles southeast of Woodside, San Mateo County, on 10 March between 9:30 and 10:30 and on 4 April 1963.

One female of *L.* (*E.*) *pullilabre* (Vachal) was found with a small pollen load from flowers of *Camissonia campestris* at 10 miles west of Simmler, San Luis Obispo County, on 3 May 1962. This bee, although polylectic, is one of the principal pollinators of *Clarkia* (MacSwain, Raven, and Thorp, 1973).

Each of the last four bee species was found once in the Great Basin studies.

A total of 134 females of *L.* (*E.*) *miguelense* (Cockerell) have been taken while gathering pollen from flowers of five species of *Camissonia* in southern cismontane California from Los Angeles and western Riverside counties north to Monterey and Madera counties as follows. Kern Co.: 18 miles east of Bakersfield on *C. campestris,* five on 11 April 1958 between 6:45 and 8:15, eight on 20 June 1958 between 6:10 and 7:30, and one on 3 April 1969 at 8:02. Los Angeles Co.: Claremont on *C. bistorta,* one on 29 April 1959 at 8:10. Madera Co.: 7 miles south of Coarsegold on *C. campestris,* one on 16 May 1963 between 8:00 and 8:15; 3.7 miles west of Oakhurst on *C. s. sierrae,* five on 15 May 1963 between 7:30 and 9:20; 1.2 miles west of Oakhurst on *C. s. sierrae,* four on 13 May 1965. Monterey Co.: 1.5 miles west of Bradley on *C. campestris,* 39 on 3 May 1963 between 8:10 and 10:15. Riverside Co.: 5 miles south of Sage on *C. bistorta,* five on 17 April 1965; 7 miles E. Temecula on *C. bistorta,* one on 23 April 1959 at 6:35. San Luis Obispo Co.: 1.8 miles west of Bee Rock on *C. campestris obispoensis,* 18 on 6 May 1963 between 6:40 and 9:30; Creston on *C. campestris,* one on 10 April 1961 between 8:10 and 8:20; 2.5 miles south of Creston on *C. campestris,* two on 11 April 1961 at 7:34 and 8:54; 4.5 miles northeast of Santa Margarita on *C. californica,* one on 28 April 1959 at 9:53, and on *C. contorta,* one on 2 May 1959 at 8:39; 5 miles northeast of Santa Margarita on *C. contorta,* one on 13 June 1963; 6.8 miles southwest of Shandon on *C. campestris,* six on 23 March 1961 between 8:40 and 10:15, one on 2 April 1961 at 7:28, two on 3 April 1961 between 7:01 and 7:06, and one on 10 April 1961; 10 miles west of Simmler, on *C. campestris,* one on 5 May 1962 between 6:50 and 7:00, one on 6 May 1962 between 6:50 and 7:00, and 16 on 13 May 1962 between 7:20 and 8:27. Santa Barbara Co.: 28 miles northwest of New Cuyama on *C. campestris,* six on 8 June 1963 between 6:50 and 7:30. Ventura Co.: Hungry Valley, 5 miles south of Gorman on *C. campestris,* four on 10 April 1960 between 7:35 and 8:54, and two on 7 May between 6:15 and 6:29.

Four females of an undescribed species related to *L.* (*E.*) *galpinsiae* (Cockerell) [determined by G. E. Bohart] were found in association with flowers of *Oenothera deltoides cognata* at 2 miles southwest of Livingston, Merced County, on 16 April 1959 between 18:25 and 18:45. Two of these females were collecting pollen. Another female without pollen was taken at the same locality and flowers on 4 April 1959 between 17:30 and 18:30. In addition 151 females and 75 males in the California Insect Survey collection were taken at Antioch, Contra Costa County, between 1936 and 1953. Females were taken between 15 March and 3 July and the males were taken between 23 May and 11 August. Most of the females (126) were taken in association with flowers of *O. deltoides howellii.* The foraging activities of the Antioch population of this crepuscular *Oenothera* oligolege are described by Turner (1966) who also records females as late as 19 September and males until 22 September. The closely related *L.* (*E.*) *galpinsiae* was encountered in each of the areas we have studied (Linsley, MacSwain, and Raven,

1963*a*, 1963*b*, and 1964) and has been associated with evening-opening species of *Camissonia, Gaura,* and *Oenothera.* Knerer and MacKay (1969) described the biology of *L.* (*E.*) *oenotherae* (Stevens) which collected pollen from onagraceous cultivars.

Lasioglossum (*Dialictus*) spp.

The subgenus *Dialictus* is represented in our collections by four determined species plus three undescribed species and several unidentified specimens for a total of 36 females and one male. These bees are principally pollen scavengers and pollen thieves since, with their small size and pollen-gathering behavior, they do not come into contact with the stigmatic surfaces of the flowers and thus do not pollinate them.

One female of *L.* (*D.*) *brunneiventre* (Crawford) was found at Hungry Valley, 5 miles south of Gorman, Ventura County, on *Camissonia campestris* on 10 April 1960 at 9:17.

One male and 12 females of *L.* (*D.*) *californiae* (Sandhouse) were taken at the following localities. Los Angeles Co.: Coldwater Canyon, San Gabriel Mts., on *Oenothera hookeri,* one male, two females on 10 August 1961 at 12:00. Monterey Co.: 17 Mile Drive on *Camissonia* sp., one female on 23 May 1959 at 9:30. Riverside Co.: 5 miles south of Temecula on *Oe. hookeri,* one female on 16 July 1961. San Bernardino Co.: East Highland on *Oe. hookeri,* one female on 27 July 1961. San Luis Obispo Co.: Bee Rock, 1.8 miles west, on *C. campestris,* two females on 6 May 1963 at 7:10 and between 8:00 and 8:05; Shandon, 5 miles west on *C. campestris,* one female 11 April 1960 at 8:06; Simmler, 10 miles west on *C. campestris,* one female on 13 May 1962 at 8:33. San Mateo Co.: 1.7 miles southeast of Woodside on *C. ovata,* one female on 10 March 1963 between 9:30 and 10:30. Ventura Co.: Hungry Valley, 5 miles south Gorman on *C. campestris,* two females on 7 May 1959 at 6:51 and 7:11.

Two females of *L.* (*D.*) *incompletum* (Crawford) were encountered, one at 1.5 miles west of Bradley, Monterey County, on *C. campestris* on 3 May between 9:50 and 10:05; and one at Rancho Santa Ana Botanic Garden, Claremont, Los Angeles County, on *Oenothera hookeri* on 9 July 1961. This species was also encountered at four localities in the Great Basin (Linsley, MacSwain, and Raven, 1963*b*).

Two females of *L.* (*D.*) *nevadense* (Crawford) were encountered on *C. campestris,* one at 1.5 miles west of Bradley on 3 May 1963 between 9:50 and 10:05, and one at 7.6 miles south of Coarsegold, Madera County, on 7 May 1963 between 8:45 and 8:51.

The aggregate records for the undescribed species and unidentified specimens of *Lasioglossum* (*Dialictus*) are as follows. Baja California, Mexico: 14 miles south of San Quintín, 4 females on *C. crassifolia* on 20 March 1962 between 8:00 and 8:35. California. Merced Co.: 2 miles southwest and 2.5 miles south of Livingston, six females on *C. campestris* and *C. contorta* in late March and early April between 8:28 and 14:30. Monterey Co.: 1.8 miles west of Bee Rock, 2 females on *C. campestris obispoensis* on 6 May 1963. Riverside Co.: 5 miles south of Sage, one female on *C. bistorta* on 17 April 1965 between 8:30 and 9:00. San Luis Obispo Co.: 10 miles west of Simmler, one female on *C. campestris* on 13 May 1962.

Agapostemon spp.

Three species of *Agapostemon* were observed gathering pollen from species of *Camissonia* and *Oenothera* in cismontane California and Baja California.

Three females of *A. femoratus* Crawford were collected gathering pollen as follows. Bakersfield, 18 miles south on *Camissonia campestris* on 3 April 1960 at 7:43; Livingston, 2.5 miles southwest on *C. campestris* on 4 April 1959 at 8:51; and Turlock, on *Oenothera hookeri* on 5 June 1959 between 6:00 and 6:15. This species (treated as *A. cockerelli* Crawford) was more abundant on *Camissonia* and *Oenothera* in the Great Basin.

Females of *A. peninsularis* Roberts were taken at two localities in northern Baja California, Mexico, as follows: 14 miles south of San Quintín, 20 (18 with pollen) on *Camissonia crassifolia* on 30 March 1962 between 7:31 and 8:35; 3 miles south of Ensenada, one with pollen on *C. cheiranthifolia suffruticosa* on 29 March 1962 at 8:05; Guerrero Negro, just at the line between the State (Norte) and the Territory (Sur), two without pollen on *C. crassifolia* on 29 March 1972 between 6:45 and 7:00.

The most abundant *Agapostemon* in the present study, *A. texanus* Cresson, is represented in our collections by 102 females and 2 males as follows. Baja California, Mexico: Santo Tomás, one female on *C. californica* on 31 March 1962 at 8:20. California. Kern Co.: Bakersfield, 18 miles east, on *C. campestris,* five females on 11 April 1958 between 6:45 and 8:15, three females on 20 March 1960 between 7:33 and 8:10, two females on 3 April 1960 at 7:14 and 7:46, two females on 1 April 1962 between 8:10 and 8:30, and one female on 2 April 1962 between 9:30 and 10:05; on *C. boothii decorticans,* two females on 27 April 1959. Los Angeles Co.: Claremont, one female on *C. bistorta* on 19 April 1959 at 10:29; Point Dume, one female on *C. cheiranthifolia suffruticosa* on 5 April 1959 at 9:22; Rancho Santa Ana Botanic Garden, Claremont, 15 females, one male on *Oenothera hookeri* on 9 July 1961, and one female on *Oe. californica* on the morning of 18 July 1961. Madera Co.: Oakhurst Cemetery, two females on *C. s. sierrae* on 16 May 1963 between 7:20 and 7:30, and 8:30 and 8:45. Merced Co.: Livingston, 2 miles southwest on *C. campestris,* two females on 4 April 1959 at 8:51 and 9:12, and one female on 5 April 1959 at 9:10; Livingston, 2.5 miles south on *C. campestris:* one female on 20 March 1960 at 9:18, one female on 6 April 1960 at 7:38, two females on 2 April 1960 at 8:02 and 8:09; and on *C. contorta:* two females on 26 March 1960 at 10:58. Monterey Co.: Bradley, 1.5 miles west on *C. campestris,* three females on 3 May 1963 between 8:40 and 10:05. Sacramento Co.: Brannan Island State Park on *C. c. cheiranthifolia,* 11 females on 20 May 1965. San Bernardino Co.: East Highland on *Oenothera hookeri,* one female on the morning of 27 July 1961 and three females on the morning of 6 August 1961. San Francisco Co.: Fleishhacker Zoo, southwest corner on *C. c. cheiranthifolia,* one female on 19 May 1959 at 9:00. San Luis Obispo Co.: Bee Rock, 1.8 miles west on *C. campestris obispoensis,* two females on 6 May 1963 between 7:30 and 9:30; Creston on *C. c. campestris,* one female on 11 April 1961 between 9:00 and 9:10; Santa Margarita, 4.5 miles east on 28 April 1959, one female on *C. contorta* at 9:52, and two females on *C. californica* at 10:03 and 10:07. Santa Barbara Co.: Pine Canyon Guard Station, 1.6 miles west, and 2.6 miles east on *C. campestris,* two females on 23 May 1959 at 7:08 and 8:07, respectively. Stanislaus Co.: Turlock on *Oe. hookeri,* 22 females on 5 June 1959 between 6:00 and 6:45. Tulare Co.: 2.4 miles south of Tipton on *C. campestris,* one female on 29 March 1960 at 9:52, and one female on 2 March 1961; Tulare Airpark on *C. campestris,* one female on 6 April 1964 between 8:35 and 8:45, and three females on 16 April 1964 between 7:20 and 7:50. Ventura Co.: Hungry Valley, 5 miles south of Gorman on *C. campestris,* one female on 7 May 1959 at 7:11; San Buena Ventura State Park, south end on *C. cheiranthifolia suffruticosa,* one male on 8 June 1959 at 6:55; Ventura on *C. c. suffruticosa,* one female on 19 April 1959 at 8:39. We also found *A. texanus* visiting both *Camissonia* and *Oenothera* in about ten localities in the Great Basin.

Augochlorella pomoniella (Cockerell)

One female was taken collecting nectar at 2.5 miles south of Creston, San Luis Obispo County, on *Camissonia campestris* at 9:26 on 11 April 1961.

Sphecodes sp.

One female of this parasitic bee was taken at 2.5 miles south of Livingston, Merced County, on *Camissonia campestris* at 14:28 on 24 March 1960.

Dufourea spp.

Four species and a few undetermined specimens of *Dufourea* (*Halictoides*) were taken in association with *Camissonia* in this study. Only two of these, *D. virgata* (Cockerell) and *D. boregoensis* (Michener), were encountered in our previous studies. All the species of *Dufourea* were encountered later in the morning than the other oligoleges of *Camissonia* and they served as scavengers of residual pollen. Due to their small size and behavior patterns in collecting pollen, they did not usually touch the stigma and therefore were ineffective as pollinators and may be considered pollen thieves.

One female of *D. boregoensis* (Michener) was taken without pollen at Hungry Valley, 5 miles south of Gorman, Ventura County, on *C. campestris* on 7 May 1959 at 7:11. A male and female were taken at 1.8 miles west of Bee Rock, San Luis Obispo County, on *C. c. obispoensis* at 11:00 on 5 May 1963 and 8:45 to 8:50 on 6 May 1963 respectively. This species was encountered on both the Colorado and Mojave Deserts on *Camissonia.*

One male of *D. mulleri* (Cockerell) was taken at 5 miles south of Sage, Riverside County, on *C. bistorta* between 8:30 and 9:00 on 17 April 1965.

The most commonly encountered species of *Dufourea* in our study was *D. scintilla* (Cockerell) represented by 48 males and 238 females (mostly collecting pollen) as follows. Los Angeles Co.: Point Dume, one female on *C. cheiranthifolia suffruticosa* on 5 April 1959 at 9:43. Madera Co.: On *C. campestris* on 7 May 1963, at Coarsegold, 3.9 miles south, seven females between 6:45 and 9:15; at Coarsegold, 7.6 miles south, one male, 13 females between 8:05 and 8:59. On *C. sierrae* at Oakhurst Cemetery, five males, 21 females on 16 May 1963 between 7:30 and 9:15;

at Oakhurst, 3.7 miles west, eight males, seven females on 15 May 1963 between 8:10 and 9:30. Monterey Co.: Bee Rock, 1.8 miles west on *C. campestris obispoensis,* three males, one female on 5 May 1963 at 11:00 and ten males, seven females on 6 May 1963 between 7:50 and 9:30. Bradley, 1.5 miles west, on *C. campestris,* three males, one female on 20 April 1963 and 11 females on 3 May 1963 between 8:50 and 10:05. Orange Co.: Newport Beach on 24 April to 4 May 1962, 20 females, most (13) with pollen from *Camissonia* sp. (probably *C. cheiranthifolia suffruticosa*). Riverside Co.: Temecula, one female on *C. contorta* on 23 April 1959 at 8:07. San Luis Obispo Co.: Creston on *C. campestris,* three females on 10 April 1961 between 8:20 and 9:10 and six females on 11 April 1961 between 7:40 and 9:20; Creston, 2.5 miles south on *C. campestris,* ten females, one male on 11 April 1961 between 7:35 and 9:25; Santa Margarita, 4.5 miles northeast on 28 April 1959, three females on *C. contorta* between 10:31 and 11:10, and one female on *C. micrantha* at 11:04; Santa Margarita, 5 miles northeast on *C. campestris,* two males and 12 females on 2 May 1962 and two females on 4 May 1962; Shandon, 6.8 miles southwest on *C. campestris,* one female on 23 March 1961 between 9:30 and 9:40, four females on 2 April 1961 between 6:15 and 7:38, five females on 3 April 1961 between 7:12 and 7:28, and 76 females on 10 April between 8:29 and 9:00. Tulare Co.: Tipton, 2.4 miles south on *C. campestris,* two females on 29 March 1960 at 10:15 and 10:24; Tulare Airpark on *C. campestris,* two females, eight males on 6 April 1964 between 8:35 and 8:55 and 15 females, one male on 16 April 1964 between 7:20 and 8:00. Tulare, 4.5 miles south on *C. campestris,* one male on 29 March 1960 between 8:30 and 8:40. Baja California, Mexico: Santo Tomás, 7 miles north, five females on *C. bistorta* on 31 March 1962 between 8:45 and 9:25; El Rosario, 5 miles east, five males, one female on *C. lewisii* on 11 March 1960 near noon.

Three females and four males of *D. truncata* Timberlake were found as follows: one female with pollen on *C. campestris* at Hungry Valley, 5 miles south of Gorman, Ventura County, on 10 April 1960 between 10:15 and 10:30; one male on *C. bistorta* at Homeland, Riverside County, on 14 March 1962 at 10:00; three males, two females with pollen on *C. bistorta* at 5 miles south of Sage, Riverside County, between 7:30 and 9:30 on 17 April 1965.

One female with pollen and 13 males of *D. virgata* (Cockerell) were encountered as follows: one female on *C. boothii* at 18 miles east of Bakersfield, Kern County, on 19 April 1958; five males on *C. campestris* at 6.8 miles southwest of Shandon, San Luis Obispo County, on 23 March 1961; two males on *C. campestris* at Hungry Valley, 5 miles south of Gorman, Ventura County, on 10 April 1960 between 8:58 and 10:30; six males sleeping on *C. campestris* at 1.6 miles west of Bradley, Monterey County, on 20 April 1963. This species was also encountered on the Colorado Desert.

Additional undetermined specimens are as follows. Merced Co.: Livingston, 2 miles southwest, one male on *C. campestris* on 4 April 1959. Ventura Co.: Hungry Valley, 5 miles south of Gorman on *C. campestris,* two females, two males on 10 April 1960 between 10:15 and 10:30.

Hesperapis spp.

Two species of *Hesperapis* (*Panurgomia*) were taken in association with *Camissonia* in this study. One may be an oligolege on *Camissonia,* while the other is an oligolege of *Clarkia.*

Hesperapis nitidula Cockerell has been associated with *Camissonia bistorta* (Stage, 1966:35, 83). In our collections it is represented by 12 females with pollen and 28 males on *Camissonia* as follows. Baja California, Mexico: Descanso Dunes, south of Ensenada on *C. bistorta,* one female, 16 males on 31 March 1962 at 15:00; San Quintín, 14 miles south on *C. crassifolia,* seven females, five males on 30 March 1962 between 7:55 and 8:35 with four of the males collected in the afternoon; Santo Tomás, 7 miles north on *C. bistorta,* two males on 31 March 1962 between 8:00 and 8:15. California. Kern Co.: Bakersfield, 18 miles east on *C. campestris,* one female on 31 March 1962, one female, one male on 1 April 1962 between 9:10 and 9:20. Madera Co.: Coarsegold, 7.6 miles south on *C. campestris,* one female on 7 May 1963 at 9:10. Monterey Co.: Bradley, 1.5 miles west on *C. campestris,* one female on 3 May 1963 between 9:50 and 10:05. Riverside Co.: 5 miles south of Sage on *C. bistorta,* 3 males between 6:45 and 9:00 on 17 April 1965. San Luis Obispo Co.: Bee Rock, 1.8 miles west on *C. campestris obispoensis,* one male on 6 May 1963 between 8:55 and 9:30. We also encountered its close relative *H. wilmettae* Cockerell on the Colorado and Mojave Deserts (Linsley, MacSwain, and Raven, 1963*a,* 1964).

Hesperapis regularis (Cresson) is an oligolege of *Clarkia* (Stage, 1966; MacSwain, Raven, and Thorp, 1973). In our collection on *Camissonia campestris* it is represented by two females without pollen and 10 males as follows. Santa Barbara Co.: Pine Canyon Guard Station, 1.6 miles west and 2.6 miles east on 23 May 1959 between 6:45 and 12:00.

Megachile spp.

Two species of *Megachile* are represented in our collections by one female each. The female of the widespread, polylectic *M. perihirta* was taken with a mixed load of pollen at Quincy, Plumas County, on *Oenothera hookeri* on 2 July 1963. This bee was also encountered on *Oenothera* in the Great Basin. The female of *M.* sp. near *addenda* Cresson was taken at Point Dume, Los Angeles County, on *C. cheiranthifolia suffruticosa* on 5 April 1959 at 10:17.

Osmia spp.

Two males of *Osmia brevis* Cresson were taken at 2.5 miles south of Livingston, Merced County, one on *C. contorta* on 24 March 1960 at 14:05 and one on *C. campestris* on 26 March 1960 at 10:24. One male of *Osmia bakeri* Sandhouse was taken on *C. campestris obispoensis* at 1.8 miles west of Bee Rock, San Luis Obispo County, on 5 May 1963 at 11:00. One female of *O.* sp. near *phellax* Sandhouse was taken at Creston, San Luis Obispo County, on *C. campestris* between 9:10 and 9:20 on 10 April 1961. Another female of this species was taken on this plant on the Mojave Desert near Lancaster.

Nomada spp.

Several unidentified specimens of *Nomada* were taken in association with *Camissonia* or at the nest sites of *Diandrena* oligolectic on this plant. The behavior of the latter are discussed by Thorp (1969). Our records are as follows. Kern Co.: 18 miles east of Bakersfield at nest entrance *Andrena macswaini,* one female on 11 March 1960 at 11:15, eight females on 20 March 1960 between 9:00 and 9:50. Merced Co.: Livingston, 2 miles southwest on *C. campestris,* two females on 5 April 1959 between 8:55 and 8:57; Livingston, 2.5 miles south on *C. contorta,* one female on 20 March 1960 at 9:54. Monterey Co.: 1.5 miles southeast of Carmel on *C. ovata,* two females on 30 March 1963 between 10:00 and 10:20. San Luis Obispo Co.: south end of Cambria, one male, one female cruising on 29 March 1963 between 10:15 and 10:30. These two specimens are conspecific with the *Nomada* recorded as a parasite of *A. chalybaea* by Thorp (1969).

Epeolus sp.

One female of an unidentified *Epeolus* was taken at 1.8 miles west of Bee Rock, San Luis Obispo County, on *C. campestris obispoensis* on 6 May 1963 between 8:55 and 9:30.

Melissodes tepida timberlakei Cockerell

Two females of this polylectic species were taken with mixed loads of pollen at Rancho Santa Ana Botanic Garden, Claremont, Los Angeles County, on *Oenothera hookeri* on 9 July 1961.

Peponapis purinosa (Say)

One female of this common squash bee which is an oligolege of *Cucurbita* was taken without pollen at Turlock, Stanislaus County, on *Oenothera hookeri* on 5 June 1959 between 6:30 and 6:45.

Synhalonia spp.

Two species of *Synhalonia* were encountered in our observations on *Camissonia.* Two females of *S. actuosa* (Cresson) were taken with pollen at 1.5 miles west of Bradley, Monterey County, on *C. campestris* on 3 May 1963 between 9:50 and 10:05. One female of *S. actuosa* was also found on *Camissonia* in the Great Basin. One male of an undescribed species near *S. actuosa* was taken at 7 miles north of Santa Tomás, Baja California, Mexico, on *C. bistorta* on 31 March 1962 between 8:30 and 8:45.

Anthophora spp.

Three species of *Anthophora* were taken at the flowers of *Camissonia.* One female of *A. crotchii* Cresson was taken with pollen at 7 miles north of Santo Tomás, Baja California, Mexico, on *C. bistorta* on 31 March 1962 between 7:45 and 8:00. One female of *A. neglecta* Timberlake and Cockerell was taken at 7.3 miles north of Bakersfield, Kern County, on *C. campestris* on 3 April 1962 between 7:30 and 7:40. This bee has also been found on flowers of *Camissonia* and *Oenothera* in the Great Basin and on the Colorado and Mojave Deserts. One female of *A. washingtoni*

Cockerell was taken at Creston, San Luis Obispo County, on *C. campestris* on 10 April 1961 between 7:40 and 7:50.

Emphoropsis spp.

Two species and several unidentified specimens of *Emphoropsis* were taken in association with *Camissonia* and *Oenothera.* Six males of *E. miserabilis* (Cresson) were taken at Fort Cronkhite, Marin County, on *C. c. cheiranthifolia* on 12 and 14 April 1959. Two females of *E. rugosissima* Cockerell were taken at Point Dume, Los Angeles County, on *C. cheiranthifolia suffruticosa* on 5 April 1959 at 9:46 and 10:42. This species was also found on *Camissonia* in the Great Basin and Mojave Desert. Undetermined specimens of *Emphoropsis* are as follows. Baja California, Mexico: 3 miles south of Ensenada on *C. cheiranthifolia suffruticosa,* two females with pollen, one male on 29 March 1962 between 6:58 and 7:32; 14 miles south of San Quintín on *C. crassifolia,* 12 females on 30 March 1962 between 7:00 and 8:30. California. Contra Costa Co.: Antioch on *Oe. deltiodes howellii,* one female on the afternoon of 9 May 1965. San Diego Co.: Carlsbad on *C. cheiranthifolia suffruticosa,* one female on 30 April 1959 between 8:10 and 8:15 and six females, 2 males on 21 April 1962 between 6:20 and 7:50.

Xeromelecta californica (Cresson)

One female of *Xeromelecta* was taken at 18 miles east of Bakersfield on *C. boothii* on 27 April 1959.

Melecta separata callura (Cockerell)

Two females of *Melecta* were taken at 18 miles east of Bakersfield, Kern County on *C. campestris* in 1962 between 9:20 and 9:30 on 1 April and between 7:47 and 7:50 on 2 April.

Xylocopa spp.

Two species of *Xylocopa* were collected gathering pollen from *Oenothera hookeri.* Records for *X. tabaniformis orpifex* Smith are as follows. Los Angeles Co.: Coldwater Canyon, San Gabriel Mountains on 10 August 1961, one female at 6:05; Rancho Santa Ana Botanic Garden, Claremont on 18 June 1961, one female at 20:25, and six females on 27 June 1961; San Antonio Canyon, San Gabriel Mountains, nine females on 25 July 1959 between 5:25 and 6:00. Riverside Co.: 5 miles south of Temecula, five females on 17 July 1961. San Bernardino Co.: East Highland, two females on 23 June 1961, and one female on 15 July 1961; Prairie Fork, San Gabriel Mountains, three females in the afternoon of 12 July 1961, six females in the morning of 13 July 1961, four females in the afternoon of 1 August 1961, and 18 females, one male on 7 September 1961 between 8:30 and 11:00. We also have an unpublished record of this bee in association with *Oe. parryi* in Washington County, Utah, in the Great Basin. Reecords for *X. varipuncta* Cresson on *Oe. hookeri* are as follows. Los Angeles Co.: Rancho Santa Ana Botanic Garden, Claremont, three females on 9 July 1961 and one female on 22 July 1961. Riverside Co.: 5 miles east of Temecula, one female on 17 July 1961. San Bernardino So.: East Highland, three females in the morning of 27 July 1961 and five females on 29 September 1961.

Ceratina sp.

Two females of an unidentified species of *Ceratina* were encountered without pollen at the flowers of *Oenothera hookeri* at Coldwater Canyon, San Gabriel Mountains, Los Angeles County, on 10 August 1961 at 12:00.

Bombus occidentalis Greene

One worker of this bumble bee was taken without pollen at Fleishhacker Zoo, southwest corner, San Francisco County, on *C. c. cheiranthifolia* on 19 May 1959 at 10:22.

Apis mellifera Linnaeus

The ubiquitous polylectic honey bee, introduced into California about 100 years ago, was frequently encountered at flowers of *Camissonia* and *Oenothera* in cismontane California in contrast to its rarity on these plants in the Mojave, Colorado, and Great Basin Deserts. The larger, more webby pollen grains of *Oenothera* were collected only with difficulty, while the smaller grains of *Camissonia* with their fewer and shorter viscin threads were handled with relative ease by the honey bee.

The numbers of honey bees collected do not reflect their actual abundance on the plants. Our samples are biased since honey bees were often ignored in preference to the native, oligolectic species and frequently only a small sample was taken to establish its presence for the record. Our collections of worker honey bees include the following. Kern Co.: Bakersfield, 18 miles east on *Camissonia boothii decorticans,* one on 2 April 1959, one on 28 March 1959, and one on 29 March 1960 at 17:19; on *C. campestris,* one on 27 March 1959 at 17:10, one on 3 April 1960 at 6:44, three on 1 April 1962 between 7:20 and 8:40. Lake Co.: Lakeport, 2 miles southwest on *C. ovata,* two on 24 April 1963. Madera Co.: Coarsegold, 3.9 miles south on *C. campestris,* one on 16 April 1963 between 6:45 and 7:45; Coarsegold, 7 miles south on *C. campestris,* four on 16 April 1963 between 8:00 and 8:15; Coarsegold, 7.6 miles south on *C. campestris,* one on 7 May 1963. Merced Co.: Delhi, 0.5 miles south on *C. campestris,* abundant on 6 April 1960 at 7:05; Livingston, 2 miles southwest on *C. campestris,* six on 4 April 1959 between 7:20 and 13:00, one on 5 April 1959 at 8:12, on *Oenothera deltoides cognata,* one on 17 April 1959 at 9:14; Livingston, 2.5 miles south on *C. campestris,* two on 20 March 1960 at 8:18 and 8:31 and one on 21 March 1960 at 8:51, on *C. contorta,* four on 20 March 1960 between 8:05 and 9:02. San Luis Obispo Co.: Santa Margarita, 4.5 miles northeast on *C. californica,* two on 24 May 1959 at 7:37. Santa Barbara Co.: Pine Canyon Guard Station, 2.6 miles east on *C. campestris,* three on 23 May 1959 between 6:58 and 7:13; Pine Canyon Guard Station, 1.6 miles west on *C. campestris,* four at 6:45. Shasta Co.: Hat Creek Post Office on *Oe. hookeri,* four on 11 August 1959. Stanislaus Co.: Turlock on *Oe. hookeri,* four on 5 June 1959 between 5:40 and 6:45. Tulare Co.: Fairview, 1 mile south on *C. campestris,* one on 30 April 1964; Tulare Airpark on *C. campestris,* one on 16 April 1964 between 8:00 and 8:15.

OTHER INSECT VISITORS OF CAMISSONIA AND OENOTHERA

Few other insect visitors have been taken in association with *Camissonia* and *Oenothera* and these are unimportant as pollinators. We have not, however, sampled extensively in the evening on the species of *Oenothera* in our area. *Oenothera hookeri* is principally pollinated by nocturnal moths (Gregory, 1963–64) and *Oe. deltoides* is visited by nocturnal moths in addition to the nocturnal bee *Lasioglossum (E.)* sp. near *galpinsiae*. Our records of other insect visitors include the following:

One *Heliosea pulchripennis* Grt., a diurnal noctuid moth, was taken at 7 miles north of Santa Tomás, Baja California, Mexico, on *C. bistorta* on 29 March 1962.

Two nitidulid beetles, *Meligethes dauricus* Mots., were taken on *Oe. hookeri* at Grass Valley, San Bernardino Mountains, San Bernardino County, California, on 31 July 1961.

One syrphid fly, *Mesograpta marginata* (Say), was taken on *C. campestris obispoensis* at 1.8 miles west of Bee Rock, San Luis Obispo County, California, on 5 May 1963.

A tachinid fly on *C. c. cheiranthifolia* at Grover City, San Luis Obispo County, California, on 28 April 1959.

One worker of *Vespula pensylvanica* (Saussure), our most common yellowjacket wasp, was taken on *Oe. hookeri,* at Prairie Fork, San Gabriel Mountains, San Bernardino County, California, on 7 September 1961.

Two female gorytine wasps, *Hoplisoides hamatus* (Handlirsch), were taken on *C. campestris obispoensis* at 1.8 miles west of Bee Rock, San Luis Obispo County, California, on 6 May 1963.

COMPARATIVE BEHAVIOR OF THE SPECIES OF ANDRENA AND THEIR HOST FLOWERS

The majority of the 18 species of *Andrena* treated are primarily or completely endemic to cismontane California and northwestern Baja California. Only *A.* (*Onagrandrena*) *deserticola* occurs principally outside the area—in the western Mojave Desert. *Andrena* (*O.*) *convallaria* contains two subspecies, one endemic to the Mojave Desert, the other restricted to our area. Five species occur primarily in our area, but also sparingly in the Mojave Desert—*A.* (*O.*) *furva, A.* (*Diandrena*) *parachalybea, A.* (*D.*) *foxii, A.* (*D.*) *sperryi,* and *A.* (*D.*) *apasta.* The remaining eleven species are endemic to cismontane California and Baja California—*A.* (*O.*) *blaisdelli, A.* (*O.*) *camissoniae, A.* (*O.*) *eulobi, A.* (*O.*) *oenotherae, A.* (*O.*) *oraria, A.* (*O*). *vespertina, A.* (*D.*) *chalybaea, A.* (*D.*) *eothina, A.* (*D.*) *anatolis, A.* (*D.*) *cyanosoma, A.* (*D.*) *macswaini.*

These 18 species constitute the largest number of oligolectic *Andrena* encountered thus far in our studies. We found three *Onagrandrena* in the Colorado Desert (1963*a*), seven *Onagrandrena* in the Great Basin (1963*b*) and eight *Onagrandrena* plus four *Diandrena* in the Mojave Desert (1964).

All 18 species of *Andrena* studied here are oligolectic on species of *Camissonia,* while none are associated with the large-flowered, evening-opening species with four-lobed stigmas (*Oenthera* sensu Raven, 1964). This contrasts with our earlier studies in which one to three of the *Onagrandrena* were associated with *Oenothera.* Most of the *Andrena* in the present area collect pollen from more than one species of *Camissonia,* but all exhibit a primary relationship with one plant species. Each of four bee species exhibits a primary association with one of the following plants: *C. boothii, C. crassifolia, C. ovata,* and *C. bistorta.* Two species of bees have primary associations with *C. cheiranthifolia,* each bee with a different subspecies of plant. Most of the bees, 12 species or 67 percent, have their primary association with *C. campestris.* This produces situations where sympatric species of oligolectic *Andrena* compete in collecting the pollen food requisite for their offspring. In addition to competition among bees with their primary pollen association on the same plant, most of the bees exhibit secondary or tertiary pollen associations with other *Camissonia* species, thus increasing the number of oligolectic bees in each competitive complex.

Andrena (*Diandrena*) *chalybaea* is endemic to the Coast Ranges of central California within the range of its sole pollen source, *Camissonia ovata.* Populations of the bee occur with the southernmost plant populations, but we have no records of *A. chalybaea* in association with this plant north of Mendocino County. The perennial *C. ovata* and its principal pollinator, *A. chalybaea,* are associated with clay soils near the coast. Most of the remaining species of *Camissonia* and their *Andrena* oligoleges are associated with deep sandy soils. The mating flight pattern of the males encompasses areas of exposed soil where females frequently nest and areas of *C. ovata* which also contain nests where the vegetation is relatively sparse. Females collect pollen for about two to four hours starting shortly after the sunlight reaches the flowers. As the pollen and nectar become depleted during the morning, the bees visit more flowers per minute, spend more time in flight searching, and spend less time on each flower. At the end of the daily foraging period they obtain additional nectar from *Ranunculus californicus.* Few other insects visit *C. ovata* flowers. Less than 10 percent of the localities of *A. chalybaea* were shared with females of *A.* (*D.*) *apasta* and *A.* (*O.*) *oraria oraria* in competition for pollen of *C. ovata* and even in these areas *A. chalybaea* remained the most abundant and dominant species.

Andrena (*D.*) *parachalybea* is most closely related to and occurs southeast of the range of *A. chalybaea* in southwestern California and northwestern Baja California. Females gather pollen from *Camissonia* flowers in the early morning. Along the coast they primarily visit *C. bistorta,* but also collect pollen from *C. cheiranthifolia suffruticosa.* In the interior valleys of the southern Coast Ranges and the southern end of the Central Valley, females gather pollen from *C. campestris.* Thus, it is allopatric from and has different pollen plant relationships from *A. chalybaea.* However, at one or more localities it is sympatric with and collects pollen from the same plants as each of the seven other *Diandrena,* as well as four species of *Onagrandrena.* In at least 25 percent of its localities it coexists with *A. sperryi, A. cyanosoma,* and *A. anatolis.* It is most abundant and dominant in the central part of its distribution.

Andrena (*D.*) *foxii* occurs in the interior valleys of southern California where its females gather pollen exclusively from *Camissonia campestris.* It is the only species of *Diandrena* with black integument on the head and thorax. Males of *A. foxii* establish "territories" among the *Camissonia* plants. Within these areas they challenge all intruders and attempt to mate with females visiting the flowers or emerging from burrows interspersed among the plants. Mating attempts at the flowers are usually unsuccessful since most of these females were mated previously when they first emerged from the ground. The species is most abundant and dominant in the southern part of the Central Valley where its females are usually the first to commence foraging each day even before females of *A.* (*O.*) *convallaria.* However, its abundance and dominance decreases in the interior valleys of the south Coast Ranges where *A.* (*D.*) *apasta* and other species predominate. It co-occurs and shares the same pollen source at one or more localities with all *Diandrena* except *A. chalybaea* and with three *Onagrandrena. Andrena foxii* shares more than 25 percent of its localities with *A.* (*O.*) *convallaria, A.* (*D.*) *apasta,* and *A.* (*D.*) *parachalybea.*

Andrena (*D.*) *sperryi* occurs in the interior valleys of southern California where its females gather pollen exclusively from *Camissonia campestris.* This species contains the smallest bees of the subgenus *Diandrena,* and the females are generally the last in the competitive complexes to initiate foraging each day. Thus they principally scavenge the residual pollen and due to their small body size rarely contact the stigma while gathering pollen from the anthers. Therefore, these bees effect virtually no pollination in *C. campestris.* The species shares the same pollen source at one or more localities with all *Diandrena* except *A. chalybaea* and with three *Onagrandrena.* At more than 25 percent of the localities where *A. sperryi* occurs we also found *A. parachalybea* and *A. foxii.*

Andrena (*D.*) *anatolis* occurs in southwestern California and northwestern Baja California where females collect pollen principally from *Camissonia bistorta* and also from *C. campestris* and *C. cheiranthifolia suffruticosa.* Bees of this species are nearly as small as those of *A. sperryi.* Thus, it is probably ineffective as a pollinator of *Camissonia.* The few time records available indicate that it forages early in the morning. Near Sage on 17 April 1965 *A.* (*D.*) *parachalybea* was the earliest and most abundant (50 percent) of the four oligolectic bee species foraging on *C. bistorta, A.* (*D.*) *anatolis* was second in abundance (34 percent) and began foraging at the same time as *A.* (*O.*) *oenotherae* which was third in abundance (11 percent), the last species to commence foraging, *A.* (*D.*) *cyanosoma* was also the least abundant (5 percent). At Newport Beach in late April and early May 1962 *A.* (*D.*) *anatolis* was more abundant during the season than any of the same three species, however the first *A.* (*D.*) *parachalybea* were taken a week earlier in the season than any of the other three. It occurs and collects pollen from the same source at one or more localities as most *Diandrena,* except *A. chalybaea* and *A. macswaini,* and as six *Onagrandrena. A. anatolis* coexists in 25 percent or more of its localities with *A.* (*D.*) *cyanosoma, A.* (*O.*) *oenotherae,* and *A* (*D.*) *parachalybea.*

Andrena (*D.*) *eothina* is most closely related to and occurs north of the distribution of *A. anatolis.* The two species occur sympatrically in Ventura County, but neither is abundant there. Females of *A. eothina* collect pollen principally from *C. campestris* and also from *C. sierrae.* They are primarily pollen scavengers, usually beginning their foraging after *A.* (*D.*) *foxii, A.* (*D.*) *apasta,* and *A.* (*O.*) *convallaria.* Due to their small size the females do not usually contact the stigma and are therefore ineffective pollinators of *Camissonia* like females of *A. sperryi* and probably *A. anatolis.* This bee is most abundant in the southern end of the Central Valley and adjacent mountain valleys. It is dominant only in association with *C. sierrae. Andrena eothina* collects pollen from the same source at one or more localities as all *Diandrena,* except *A. chalybaea,* and with three *Onagran-*

drena. It coexists in more than 25 percent of its localities with *A.* (*O.*) *convallaria, A.* (*D.*) *foxii, A.* (*D.*) *apasta,* and *A.* (*D.*) *cyanosoma.*

Andrena (*D.*) *cyanosoma* occurs from southern California to northwestern Baja California and gathers pollen from several species of *Camissonia,* but primarily from *C. campestris.* Females commence foraging later in the morning than most competing *Camissonia* oligoleges, therefore, it never appears as a dominant or relatively abundant species. We encountered the largest populations of this bee in the southern end of the Central Valley. At one or more localities it collects pollen from the same source as all *Diandrena,* except *A. chalybaea,* and with four *Onagrandrena. Andrena cyanosoma* shares more than 25 percent of its localities with *A.* (*D.*) *anatolis, A.* (*O.*) *convallaria, A.* (*D.*) *parachalybea,* and *A.* (*O.*) *oenotherae.*

Andrena (*D.*) *macswaini* is known from only a few localities in the southern end of the Central Valley and adjacent Sierran foothills. It is apparently restricted to deep sandy soils because of the relatively great depth of its nests when compared to those of other *Diandrena.* The mating flight pattern of the males is concentrated around the entrances of the aggregated nests. Females gather pollen only from *C. campestris* and *C. sierrae.* They are among the last of the competing *Camissonia* oligoleges to commence foraging each morning although they are larger than those of other late-morning, principally pollen-scavenging *Diandrena. Andrena macswaini* shares its pollen source at one or more localities with all *Diandrena,* except *A. chalybaea* and *A. anatolis,* and with two *Onagrandrena.* It shares more than 25 percent of its localities with *A.* (*O.*) *convallaria, A.* (*D.*) *eothina, A.* (*D.*) *foxii, A.* (*D.*) *cyanosoma,* and *A.* (*D.*) *sperryi.*

Andrena (*D.*) *apasta* occurs in the valleys of the Coast and Transverse Ranges from Marin County to Los Angeles County. Females gather pollen principally from *C. campestris,* but occasionally from *C. ovata* in the northernmost localities of the bee's occurrence. It is most abundant in the valleys of the south Coast Ranges where it is a dominant *Camissonia* oligolege and one of the earliest bees to start foraging in the morning. *Andrena apasta* coexists and shares its pollen source at one or more localities with all *Diandrena* and three *Onagrandrena.* It shares more than 25 percent of its localities with *A.* (*O.*) *convallaria, A.* (*D.*) *eothina, A.* (*D.*) *foxii,* and *A.* (*D.*) *sperryi.*

Andrena (*Onagrandrena*) *blaisdelli* occurs in the valleys of the mountain ranges in southern California. Females collect pollen principally from *C. campestris,* and occasionally from other species of *Camissonia* and is one of the earliest species to begin foraging in the morning. *Andrena blaisdelli* shares its pollen source at one or more localities with all *Diandrena,* except *A. chalybea* and *A. macswaini,* and with four *Onagrandrena.* It shares more than 25 percent of its localities with *A.* (*D.*) *cyanosoma, A.* (*D.*) *anatolis, A.* (*D.*) *parachalybea,* and *A.* (*O.*) *oenotherae.*

Andrena (*O.*) *camissoniae* is known only from the type locality in the south end of the south Coast Ranges, where it collects pollen from *Camissonia campestris.* It competes with *A.* (*O.*) *convallaria* for pollen at this locality.

Andrena (*O.*) *convallaria convallaria* occurs throughout the southern half of the Central Valley and the valleys of the south Coast Ranges. Throughout its range it is one of the most abundant and dominant of the *Camissonia* oligoleges. Females gather pollen principally from *C. campestris* and are usually among the first to commence pollen foraging each morning. Their large size and pollen-gathering behavior insures maximum contact with the stigma, thus these bees are extremely efficient *Camissonia* pollinators. *Andrena convallaria* coexists and shares its pollen source at one or more localities with all *Diandrena,* except *A. chalybaea,* and with four *Onagrandrena.* It shares 25 percent or more of its localities with *A. eothina, A. foxii, A. cyanosoma, A. apasta,* and *A. parachalybea.*

Andrena (*O.*) *deserticola* enters our area only at two localities: Hungry Valley, and 18 miles west of Bakersfield although its principal distribution is along the western edge of the Mojave Desert. The females gather pollen from *C. campestris* early in the morning. They are very rare in our area and encounter competition for pollen from all *Diandrena,* except *A. chalybaea* and *A. cyanosoma,* and from *A.* (*O.*) *blaisdelli* and *A.* (*O.*) *convallaria.*

Andrena (*O.*) *eulobi* is known from San Quintín and Guerrero Negro, Baja California, where it collects pollen from *C. crassifolia.* Since these are in the northern part of the plant's distribution, the range of the bee probably extends further south. At the type locality it co-occurs with *A.* (*D.*) *anatolis.*

Andrena (*O.*) *furva* occurs in only four localities in our area—all in valleys in the mountain ranges of southern California. Its principal pollen source is apparently *C. campestris,* but we also found it occasionally collecting pollen from *Clarkia cylindrica. Andrena furva* is a rare species in our area most frequently competing with *A.* (*D.*) *anatolis, A.* (*D.*) *cyanosoma, A.* (*O.*) *blaisdelli,* and *A.* (*O.*) *convallaria.*

Andrena (*O.*) *oenotherae* occurs along the coast of northern Baja California and southern California and extends into the Los Angeles Basin and interior valleys of the Peninsular Ranges. Females collect pollen from *C. cheiranthifolia suffruticosa* coastally and *C. bistorta* interiorly. It is an abundant and dominant bee throughout most of its range, and the females are usually the very earliest to commence foraging each morning. It competes for pollen with *A.* (*D.*) *anatolis* and *A.* (*D.*) *cyanosoma* in more than 25 percent of its localities and less frequently with *A.* (*D.*) *parachalybea* and *A.* (*O.*) *blaisdelli.*

Andrena (*O.*) *oraria* contains two subspecies restricted to short stretches of coast north and south of San Francisco Bay. Females of both collect pollen from *C. c. cheiranthifolia,* while a few females of *A. o. oraria* were encountered gathering pollen from *C. ovata* in competition with *A.* (*D.*) *chalybaea.*

Andrena (*O.*) *vespertina* is known only from two widely separated localities (fig. 15) where it collects pollen in the late afternoon from *C. boothii decorticans.* At the sites near Bakersfield it coexists with other *Camissonia* oligoleges, but does not enter into competition for pollen with them since it forages on a different host and at a different time of day.

We have indicated the number of species which coexist and utilize the same pollen source. Each of four species: *A.* (*O.*) *convallaria, A.* (*D.*) *eothina, A.* (*D.*) *foxii, and A.* (*D.*) *apasta,* commonly co-occur with each other in 21 to 43 percent of their combined localities. All four are present near Shandon and Simmler, however *A. apasta* is rare near Bakersfield. *A. foxii* is rare near Shandon, and *A. eothina* is rare near Simmler. The greatest overlap is between *A. convallaria–A.*

eothina, and *A. convallaria–A. foxii.* Other pairs of species coexisting at more than 20 percent of their combined localities include *A. cyanosoma–A. anatolis,* and *A. sperryi–A. foxii.*

We have also indicated that 12 of 13 species have as their primary pollen source *Camissonia campestris.* The remaining species, *A.* (*D.*) *anatolis,* is only secondarily associated with this plant since this bee occurs at the periphery of the plant's distribution and is primarily associated with *C. bistorta.* Seven of the 12, *A.* (*D.*) *parachalybea, A.* (*D.*) *eothina, A.* (*D.*) *cyanosoma, A.* (*D.*) *macswaini, A.* (*D.*) *apasta, A.* (*O.*) *blaisdelli,* and *A.* (*O.*) *convallaria,* also utilize other species of *Camissonia* as secondary pollen sources when the bees occur outside the range of *C. campestris,* and enter into competition with different complexes of bee species in these situations.

The fact that members of two subgenera of even a large genus like *Andrena* can coexist when they are dependent on the same pollen source is in itself of biological interest. But here we have several situations in which members of two or more species within a single subgenus, either *Diandrena* or *Onagrandrena,* coexist and utilize the same pollen source!

Explanations as to how these situations of direct competition among related bee species might arise and be maintained have been set forth by Linsley (1961) and Thorp (1969). As characteristic of bees which exhibit group oligolecty Linsley (1961) suggests that "when sympatric siblings are in direct competition with one another for pollen, exhibiting only slight differences in flower ethology, one can assume they rose allopatrically and have relatively recently come together." Thorp (1969) suggests that such complexes of related species competing for the same pollen source are permitted and maintained by one or more of the following: (1) The most closely related taxa are not involved, but tend to be allopatric or if sympatric have different pollen sources and may forage at a different time of day. Therefore the competitive complexes consist of species from a different phyletic lines even though they are consubgeneric. (2) The bees subdivide, both diurnally and seasonally, the relatively short pollen-collecting period. The former has been especially well illustrated in the present as well as our earlier papers in this series. The latter, while more difficult to document is indicated by our observations during different parts of the bloom period at single localities over several years. In both types of samples the peaks of population activity are often separate despite considerable overlap in overall activity periods. (3) Correlated with diurnal division of the pollen source are tendencies for the females of the earliest foraging species to be larger and darker than those of the latest species to become active. (4) There are differences between localities and sometimes between seasons within localities in species composition and population sizes in the competitive complexes. Among the species with larger females which forage earliest in the morning, one may be dominant in some areas while in others it may exist at low population levels or reinvade each year from areas of greater abundance.

Distribution patterns of species of *Camissonia* and their *Andrena* oligoleges are usually not entirely congruent. Those bee species restricted to a single pollen source throughout their range (e.g., *A.* (*D.*) *chalybaea* on *C. ovata; A.* (*D.*) *foxii, A.* (*D.*) *sperryi, A.* (*O.*) *camissoniae,* and *A.* (*O.*) *deserticola* on *C. campestris; A.* (*O.*)

eulobi on *C. crassifolia;* and *A.* (*O.*) *vespertina* on *C. boothii decorticans*) tend to fall short of the entire range of their pollen source. Most of the remaining bee species do not occur throughout the range of their primary pollen source and are found outside its range in association with varying portions of the ranges of other species of *Camissonia* which serve as secondary or alternative pollen sources.

Man has exerted considerable influence, especially of a disruptive nature, on the distribution patterns of the *Camissonia* and their oligolectic *Andrena* in cismontane California. The plants are not weedy types which readily adapt to distribution habitats and the bees tend to construct shallow nests, less than 30 cm deep. Thus many of man's activities tend to destroy and fractionate their populations. Many of our study sites succumb to the plow or bulldozer after leveling for roads, cultivation, or urbanization. This type of change was not encountered to so great an extent in our previous studies (Linsley, MacSwain, and Raven, 1963*a*, 1963*b*, 1964) since they did not include such extensive agricultural or urban areas.

EVOLUTION IN CAMISSONIA

Along the western edge of the Mojave Desert, we found that the pollination systems of *Camissonia kernensis* subsp. *kernensis* and *C. campestris* differed modally (Linsley, MacSwain, and Raven, 1964). The principal difference lay in the earlier opening time of *C. kernensis* and its role as the main pollen source of the large bee *Andrena mojavensis* Linsley and MacSwain, which is active very early in the morning. It appears possible that the early opening of *C. kernensis* here is an adaptive response, as this species is apparently incapable of hybridizing with *C. campestris.* It would therefore be advantageous, given the presence of an early-active bee, for the flowering times of the two plants to diverge. In apparent contradiction to such a view is the observation that where it grows sympatrically with *C. k. kernensis, C. campestris* likewise opens its flowers earlier than normal (Linsley, MacSwain, and Raven, 1964).

At any rate, it appears highly unlikely that changes in pollination system played any role in the initial divergence of these two rather distantly related species. They share the same spectrum of pollinators now to a large extent, and there appears to have been little basis for such a pattern of differentiation. Furthermore, as we pointed out in discussing our studies of *Clarkia* (MacSwain, Raven, and Thorp, 1973), local taxa never seem to have a distinctive pollinator; they presumably cannot afford to! How then could such a pollinator, *unless it were shared with another, less closely related plant also found in the area of differentiation, promote the initial divergence of a distinctive new population of plants?*

The most important single element in the evolution of *Camissonia* has been the switch from outcrossing to autogamy (Raven, 1969). Just as we pointed out for *Clarkia* in the fourth paper of this series (MacSwain, Raven, and Thorp, 1973), autogamous species are rarely visited by bees , and the bees that do visit them are merely an attenuated sample of those found abundantly associated with the outcrossing species of the same plant groups. Thus changes in pollination systems *per se* do not play a role in patterns of differentiation of this sort.

Major changes in pollination systems, as from bee to noctuid, hawkmoth, or bird pollination, were discussed on pp. 4 6. These have played an obvious role in the evolution of *Camissonia,* although only three (possibly four) such shifts in pollination systems would be sufficient to account for the entire pattern of diversity in the genus. To summarize the discussion on pp. 4–6, there was a shift from bee-pollination to pollination by medium-sized moths in the common ancestor of *Camissonia* sect. *Eremothera* and another such shift within the species *C. claviformis* (sect. *Chylismia*). The common ancestor of sect. *Lignothera* was adapted to hawkmoth pollination and derived from bee-pollinated ancestors. Finally the ancestor of *C. pterosperma* (the only species of sect. *Chylismiella*) may have been fly-pollinated and derived from bee-pollinated ancestors.

Within the bee-pollinated, outcrossing species of *Camissonia* discussed in detail in this paper—namely, *C. crassifolia, C. ovata, C. cheiranthifolia, C. bistorta, C. campestris,* and *C. sierrae* (these are the only species of the genus regularly outcrossed by bees in the area we studied)—there is no evidence that changes in pollination systems have played any role whatever in evolution. All of the species are ecologically and/or geographically separated, and so, to a large extent, are their pollinators. *Camissonia cheiranthifolia* and *C. bistorta* hybridize where they come into contact, and their pollinators change gradually in a way that is not related to the distinction between these two entities. *Camissonia ovata* and *C. campestris* have modally different pollinators, but they occur, respectively, near the coast in heavy clay soil and away from the coast in light sandy soil. Where their ranges approach one another, as when *C. ovata* grows in the dry inner south Coast Ranges of Santa Clara County, it is visited by both its own principal pollinator, *Andrena chalybaea,* and a bee that is normally associated with *C. campestris, A. apasta.* The same pattern is repeated over and over and suggests strongly that bee pollination has played a very minor role in the evolution of *Camissonia,* despite the high degree of specialization and oligolecty of the bees that visit plants of this genus. This contrasts with the situation in another genus of Onagraceae tribe Onagreae, *Clarkia* (MacSwain, Raven, and Thorp, 1973), where shifts in emphasis within bee-pollinated systems, correlated with differences in petal marking and conformation of the species, appears to have played a moderately important evolutionary role.

SUMMARY

Of the 72 identified species of bees associated with *Camissonia* and *Oenothera* in cismontane California and northwestern Baja California, Mexico, 50 are polylectic, parasitoids, or rarely visit these flowers and then only for nectar. The polylectic species whose females were most commonly encountered collecting pollen from these Onagraceae include: *Lasioglossum* (*E.*) *miguelense* (134); *Agapostemon peninsularis* (23); *A. texanus* (102); *Xylocopa tabaniformis orpifex* (56); *Apis mellifera* (50+). The remaining 22 species are known or presumed to be oligolectic on *Camissonia* and *Oenothera.* Most of these are represented by 20 or more females, only four—*Dufourea boregoensis, Hesperapis nitidula, Andrena* (*O.*) *camissoniae,* and *A.* (*O.*) *deserticola*—are represented by fewer than 15 females. The relationships of the 18 most abundant oligoleges with the plant species concerned may be summarized as follows.

The large-flowered oenotheras with deeply 4-lobed stigmas—*Oe. californica, Oe.*

deltoides, and Oe. hookeri—all open their flowers near sunset and are primarily pollinated by hawkmoths (Sphingidae). In addition females of the crepuscular bee, *Lasioglossum* (*E.*) sp. near *galpinsiae,* collect pollen in the evening from flowers of *Oe. deltoides cognata* and *Oe. deltoides howellii.*

Of the eleven species of *Camissonia* in our area, only *C. boothii decorticans* is white-flowered and afternoon-opening. It is pollinated in the afternoon by females of its oligolege, *Andrena* (*O.*) *vespertina.* In localities where the bee is rare or absent it may also be pollinated by moths in the evening.

The remaining ten species of *Camissonia* are yellow-flowered and morning-opening. Four—*C. californica, C. contorta, C. graciliflora,* and *C. micrantha*—are small-flowered, autogamous species, and rarely visited by bees except where they are growing intermixed with larger-flowered species. The remaining six outcrossing species have one or more species of oligolectic bees closely associated with them.

Camissonia crassifolia was examined at the northern part of its distribution near San Quintín and Guerrero Negro, Baja California, where it was associated with populations of its own oligolectic *Onagrandrena, A. eulobi,* and one nectar-gathering female of *A.* (*D.*) *anatolis.*

Camissonia ovata is a clumped perennial of clay soils near the coast where it is primarily associated with the oligolectic *Andrena* (*D.*) *chalybaea,* except in its northernmost populations. Two other oligoleges, *A.* (*D.*) *apasta* and *A.* (*O.*) *oraria oraria,* utilize this plant as a secondary source of pollen.

Camissonia cheiranthifolia, a perennial species confined to sand dunes along the ocean, consists of two subspecies, and contains self-incompatible, self-compatible but outcrossing, and autogamous races. The outcrossing *C. c. suffruticosa* of coastal southern California and northwestern Baja California is primarily associated with *Andrena* (*O.*) *oenotherae* and with additional oligolectic bees—*A.* (*D.*) *parachalybea, A.* (*D.*) *anatolis,* and *A.* (*D.*) *cyanosoma*—which utilize this plant as a secondary pollen source. The more northern, principally autogamous, *C. c. cheiranthifolia* exhibits considerable outcrossing in the San Francisco Bay region, the few inland populations on dunes along the Sacramento River, and on Santa Cruz Island where it is associated with populations of *A.* (*O.*) *oraria, A.* (*O.*) *convallaria,* and *A.* (*O.*) *oenotherae,* respectively. It is primarily self-pollinating elsewhere, including localities where it is visited by the less abundant, less constant polylectic bees, such as *Hylaeus* sp., *Halictus tripartitus, Lasioglossum* (*L.*) *pavonotum,* and *L.* (*Evylaeus*) *kincaidii.*

Camissonia bistorta, an annual species, is related to *C. cheiranthifolia suffruticosa,* ranges inland from populations of the latter, and is often visited by the same species of oligolectic bees. The principal oligolege of *C. bistorta* is *Andrena* (*D.*) *anatolis.* Other oligolectic bees utilizing it as a secondary pollen source are *A.* (*D.*) *parachalybea, A.* (*D.*) *cyanosoma, A.* (*O.*) *blaisdelli,* and *A.* (*O.*) *oenotherae.*

Camissonia sierrae sierrae of the central Sierra Nevada serves as a secondary source of pollen for three species of *Diandrena: A. eothina, A. cyanosoma,* and *A. macswaini.*

Camissonia campestris, an annual related to *C. sierrae,* occurs throughout the southern half of the Central Valley and adjacent foothills where it is associated with the largest complex of oligolectic bees yet found with one species of Onagraceae. Twelve species of *Andrena* utilize this plant as their primary, if not sole, source of pollen—*A.* (*D.*) *parachalybea, A.* (*D.*) *eothina, A.* (*D.*) *foxii, A.* (*D.*) *sperryi, A.* (*D.*) *cyanosoma, A.* (*D.*) *macswaini, A.* (*D.*) *apasta, A.* (*O.*) *blaisdelli, A.* (*O.*) *camissoniae, A.* (*O.*) *convallaria, A.* (*O.*) *deserticola,* and *A.* (*O.*) *furva*—a thirteenth, *A.* (*D*) *anatolis,* utilizes it as a secondary pollen source. *Dufourea scintilla* is another oligolege utilizing *C. campestris* as a primary pollen source. The means by which such competitive complexes are able to coexist has received principal attention.

TABLES

TABLE 1

FLOWER VISITING ACTIVITIES OF FEMALE BEES ON CAMISSONIA CAMPESTRIS AT THREE DATES 18 MILES WEST OF BAKERSFIELD, KERN COUNTY, CALIFORNIA, 1962 (Observations on 31 March—sunrise 6:02, sky clear; 1 April—Sunrise 6:00, sky hazy; and 2 April—sunrise 5:59, sky cloudy)[a]

Species and date	Time periods							
	6:51–7.00	7:00–7:10	7:10–7:20	7:20–7:30	7:30–7:40	7:40–7:50	7:50–8:00	8:00–8:10
31 March	12°C		13°C			14.5°C		15°C
A. foxii	14P	39P	27P	40P 3N	27P 6N	20P 21N	11P 15N	2P 29N
A. convallaria				1P	5P	1P 2N	3P	2P 1N
A. eothina				2P	5P	8P	8P	26P
A. macswaini							1P	4P
Apis								
Agapostemon								
1 April		13.5°C	14.5°C		15.5°C			16°C
A. foxii		6P 1N	25P 4N	22P	19P 5N	12P 10N	10P 11N	5P 9N
A. convallaria			1N	2P	2P	3P	1P	4P
A. eothina						2P 1N	1P	1P
A. macswaini								3P
A. deserticola					1P			
Apis				1P				
2 April	13°C		13.5°C		14.5°C	15°C		14.5°C
A. foxii						6P 1N	8P 1N	11P
A. convallaria							1P	2P
A. eothina								
A. macswaini								
A. deserticola							1P	
A. parachalybea								
Apis								
Agapostemon								

[a]P = with pollen; N = without pollen; blank = no bees collected; dash = not recorded.

TABLE 1 (continued)

Time periods									Total
8:10–8:20	8:20–8:30	8:30–8:40	8:40–8:50	8:50–9:00	9:00–9:10	9:10–9:20	9:20–9:30	9:30–10:00	
	15.5°C								
2P 5N	2P 4N					—	—	—	267
1P						—	—	—	16
6P	6P					—	—	—	61
1P	4P	←——— 2P ———→				—	—	—	12
1P						—	—	—	1
1P		←——— 2P ———→				—	—	—	3
		16°C			16°C				
4P 8N	3P	1P 1N	1N	1N				—	158
	1P	1P 1N	1P					—	17
5P	6P	4P 1N	2P	1P	1P	2P 1N	3P	—	31
4P 1N	13P 1N	13P 4N	11P	7P 1N	9P	3P 3N	3P 3N	—	79
								—	1
		1P						—	2
		14.5°C			15.5°C			16.5°C	
9P 4N	7P 1N	4P 2N	6P 6N	10P 5N	6P 4N	5N	2P 2N		100
1P	1P	1P	2P			1P	2P	1P 1N	13
	1P	2P	7P	2P	4P	3P	5P	1P 2N	27
				1P 1N	3P 1N	3P	2P	14P 7N	32
									1
					1P				1
						1P		1P	2
								1P	1

TABLE 2

FLOWER-VISITING HABITS OF ANDRENA FOXII ON CAMISSONIA CAMPESTRIS 18 MILES EAST OF BAKERSFIELD, KERN COUNTY, CALIFORNIA, 31 MARCH 1962
(Each count is for a one-minute period—sunrise 6:02)

Time	No. flowers visited	No. pollen collected
6:52	5	5
6:57	6	5
7:02	6	6
7:02	6	5
7:06	7	5
7:07	7	5
7:12	7	5
7:15	9	7
7:17	11	6
7:18	10	7
7:24	10	6
7:30	11	4
7:31	5	4
7:33	8	6
7:45	7	6
7:53	13	2
7:54	8	4

TABLE 3

Flower-visiting Activity of Females of Four Species of Andrena on Camissonia campestris at Four Dates, 18 Miles East of Bakersfield, Kern County, California, 1960

Species	Time periods									
	6:15–6:30	6:30–6:45	6:45–7:00	7:00–7:15	7:15–7:30	7:30–7:45	7:45–8:00	8:00–8:15	8:15–8:30	8:30–11:00
11 March (Sunrise 6:22)										
foxii					4P 2N					
convallaria									1P	
eothina							1P	3P 1N	2N	2N
macswaini										1N
20 March (Sunrise 6:15)										
foxii			4P	5P	6N	2N	3N	1N		
convallaria					1P	2P	1N			
eothina				3P	3P	2P	1P			
macswaini					2P 1N	2P 3N	3N			
3 April (Sunrise 5:58)										
foxii	3P	1P	2P 1N	1P	2P					
convallaria		3P 1N								
eothina					1P	2P				
macswaini		1N	7P	5P	6P	4P	3P	2P 1N		
9 April (Sunrise 5:50)										
foxii		1P	4P	1N		1P 1N				
convallaria		1P 1N		1P		1P				
eothina			1P							
macswaini				6P	6P	6P 3N				

TABLE 4

Flower-visiting Activity of Females of Five Species of Andrena on Camissonia campestris, at Five Dates, 6.8 Miles Southwest of Shandon, San Luis Obispo County, California, 1961

Species	Time periods									
	6:20–6:30	6:30–6:40	6:40–7:00	7:00–7:20	7:20–7:40	7:40–8:00	8:00–8:30	8:30–9:00	9:00–9:30	9:30–10:30
8 March										
(Start of count 8:30 a.m.)										
apasta								6P 3N	2N	1P 1N
eothina								1N	2P 2N	3N
23 March										
(Sun on flowers 8:05 a.m.)										
apasta							9P 1N	5P 1N	3P 2N	2P 1N
eothina							8P 3N	21P 4N	14P 3N	12P 7N
convallaria							2P	3 P	2P 1N	
foxii							1N			
parachalybea								1N		
2 and 3 April										
(Sun on flowers 6:17 a.m.)										
apasta	3P	4P	17P	17P	7P 1N	1P				
eothina			1P	5P	3P	2P 3N	1P			
convallaria		1P	3P	2P						
parachalybea				1P						
10 April										
(Sun on flowers 8:24 a.m.)										
apasta						2P	5P 1N			
eothina							1P			

TABLE 5

FLOWER-VISITING ACTIVITY OF FEMALES OF THREE SPECIES OF ANDRENA ON CAMISSONIA CAMPESTRIS, CRESTON, SAN LUIS OBISPO COUNTY, CALIFORNIA, 10 APRIL 1961
(Start of count 7:20)

Species	Time periods									
	7:20–7:30	7:30–7:40	7:40–7:50	7:50–8:00	8:00–8:20	8:20–8:40	8:40–9:00	9:00–9:20	9:20–9:40	9:40–9:50
convallaria	2P	6P	4P	3P	1P 1N	1P	1P 1N	4N	1N	2N
eothina		1P	2P	1P	4P 1N	4P 2N	6P 3N	7P 4N	1N	4N
cyanosoma					1P 1N	1N		1P		

TABLE 6

FLOWER-VISITING HABITS OF FEMALES OF SEVEN SPECIES OF ANDRENA ON CAMISSONIA CAMPESTRIS 10 MILES WEST OF SIMMLER, SAN LUIS OBISPO COUNTY CALIFORNIA, 6 MAY 1962
(Sun on flowers at 5:55—temperature range through count 7°C–20°C)

Species	Time periods											
	6:06–6:10	6:10–6:20	6:20–6:30	6:30–6:40	6:40–6:50	6:50–7:00	7:00–7:10	7:10–7:20	7:20–7:30	7:30–7:40	7:40–7:50	7:50–8:00
apasta	1P	1P	1P	1P	1P 2N	4P	2P 2N	1P		3P	1P	1P
blaisdelli	1P	3P	4P	9P	6P 1N	6P	4P 1N	2P	2P		1P	
convallaria		2P	6P	5P	7P 2N	4P	5P	8P	5P	2P 1N	2P	1P 2N
foxii			1P 1N	1P	1N			1P 1N				
parachalybea			1P	1P			5P	2P	3P 1N	3P	2P 3N	1P 2N
eothina								1P				1P
sperryi											1P	3P

TABLE 7

FLOWER-VISITING ACTIVITY OF FEMALES OF FIVE SPECIES OF ANDRENA ON CAMISSONIA CAMPESTRIS, 2.4 MILES SOUTH OF TIPTON, TULARE COUNTY, CALIFORNIA, 29 MARCH 1960
(Sunrise 5:47—temperature range through count 9°C–16.5°C)

Species	Time periods							
	7:35–7:45	7:45–8:00		9:35–9:45	9:45–10:00	10:00–10:15	10:15–10:30	10:30–10:48
foxii..........	3N	2P 4N		1N	1N			
convallaria....				1P	1P 1N	1N	1N	1P
macswaini.....				1P 1N	2P 3N	6N	1P 6N	3N
cyanosoma....				1P	1P	3P	1P 3N	1P 3N
eothina.......				2P		2N	1N	

TABLE 8

Flower-visiting Activity of Females of Five Species of Andrena on Camissonia campestris at Hungry Valley, 5 Miles South of Gorman, Ventura County, California
(Sunrise: 10 April 1960 = 5:34; 7 May 1959 = 5:03)

Species	Time periods										
	5:43–6:00	6:00–6:15	6:15–6:30	6:30–6:45	6:45–7:00	7:00–7:30	7:30–8:00	8:00–8:30	8:30–9:00	9:00–9:30	9:30–10:30
10 April 1960											
convallaria				1P		2P	2P				
blaisdelli					3P	3P	2P				
parachalybea							2P	2P	1P	1P 1N	
sperryi								1P	1P	1N	3N
7 May 1959											
convallaria	2P	9P	5P	3P	1P	1N					
blaisdelli	1P	6P		4P	2P	1P 1N					
foxii	3P			1P							
parachalybea		2P	2P	5P	3P	1P 1N					
sperryi					2P	1P					

TABLE 9

FLOWER-VISITING ACTIVITY OF FEMALES OF TWO SPECIES OF ANDRENA ON CAMISSONIA CHEIRANTHIFOLIA SUFFRUTICOSA AT POINT DUME, LOS ANGELES COUNTY, CALIFORNIA, 5 APRIL 1959
(Sunrise 5:36)

Species	Time periods										
	6:57–7:15	7:15–7:30	7:30–7:45	7:45–8:00	8:00–8:15	8:15–8:30	8:30–8:45	8:45–9:00	9:00–9:30	9:30–10:00	10:00–10:42
oenotherae	11P	15P	10P	5P	9P	6P	5P	9P	6P	3P	2P
parachalybea		2P	1P	1P	2P	1P	5P	3P	5P	11P	17P

TABLE 10

FLOWER-VISITING ACTIVITY OF FEMALES OF ANDRENA VESPERTINA ON CAMISSONIA BOOTHII DECORTICANS AT TWO DATES, 18 MILES EAST OF BAKERSFIELD, KERN COUNTY, CALIFORNIA
(Sunset: 19 April 1958 = 18:29; 29 March 1960 = 18:15)

Date	Time periods									
	16:27–16:30	16:30–16:40	16:40–16:50	16:50–17:00	17:00–17:10	17:10–17:20	17:20–17:30	17:30–17:40	17:40–17:50	17:50–18:02
19 April 1958					1N	3P 3N	1P 3N	3P 3N	2N	2N
29 March 1960	1N	1P	1P 3N	2P 2N	2P	1P 3N				

LITERATURE CITED

COCKERELL, T. D. A.

1924. Bees of the genus *Andrena* in the collection of the California Academy of Sciences. Pan-Pac. Entomol. 1(2):57–65.

1941. Observations on plants and insects in northwestern Baja California, Mexico, with descriptions of new bees. Trans. San Diego Soc. Natur. Hist. 9(31):337–352.

GREGORY, D. P.

1963–64. Hawkmoth pollination in the genus *Oenothera*. Aliso 5(3):357–384; 5(4):385–419.

KNERER, G. and P. MACKAY

1969. Bionomic notes on the solitary *Evylaeus oenotherae* (Stevens) (Hymenoptera: Halictidae), a matinal summer bee visiting cultivated Onagraceae. Can. J. Zool. 47(3):289–294.

LINSLEY, E. G.

1938. Studies in the Andrenidae of North America. I. Proc. Calif. Acad. Sci. 4th Ser. 23(18):263–282.

1960. A new species of *Diandrena* associated with *Oenothera* in California (Hymenoptera: Andrenidae). Pan-Pac. Entomol. 36(2):97–98.

1961. The role of flower specificity in the evolution of solitary bees. Proc. 11th Int. Congr. Entomol., Vienna (1960) 1:593–596.

LINSLEY, E. G. and J. W. MACSWAIN

1955. The North American andrenine bees of the subgenus *Melandrena* with descriptions of new species. Pan-Pac. Entomol. 31(4):163–172.

LINSLEY, E. G. and J. W. MACSWAIN

1956. Further notes on the taxonomy and biology of the andrenine bees associated with *Oenothera*. Pan-Pac. Entomol. 32(3):111–121.

1961. Descriptions of *Onagrandrena* associated with *Oenothera* and *Clarkia* with taxonomic notes on other species. Pan-Pac. Entomol. 37(2):117–130.

1963. Descriptions of new species of *Onagrandrena*, principally of the *Andrena oenotherae* complex. Pan-Pac. Entomol. 39(3):189–198.

1968. A new species of *Onagrandrena* associated with *Camissonia campestris*. Pan-Pac. Entomol. 44(2):144–145.

LINSLEY, E. G., J. W. MACSWAIN, and P. H. RAVEN

1963*a*. Comparative behavior of bees and Onagraceae, I. *Oenothera* bees of the Colorado Desert. Univ. Calif. Publ. Entomol. 33(1):1–24.

1963*b*. Comparative behavior of bees and Onagraceae. II. *Oenothera* bees of the Great Basin. Univ. Calif. Publ. Entomol. 33(1):25–58.

1964. Comparative behavior of bees and Onagraceae. III. *Oenothera* bees of the Mojave Desert, California. Univ. Calif. Publ. Entomol. 33(2):59–98.

LINSLEY, E. G., J. W. MACSWAIN, and R. F. SMITH

1955. Observations on the nesting habits and flower relationships of some species of *Melandrena*. Pan-Pac. Entomol. 31(4):173–185.

MACSWAIN, J. W., P. H. RAVEN, and R. W. THORP

1973. Comparative behavior of bees and Onagraceae. IV. *Clarkia* bees. Univ. Calif. Publ. Entomol. 70:1–84.

RAVEN, P. H.

1964. The generic subdivision of Onagraceae, tribe Onagrae. Brittonia, 16(3):276–288.

RAVEN, P. H.

1969. A revision of the genus *Camissonia* (Onagraceae). Contrib. U. S. Nat. Herb. 37(5):161–396.

STAGE, G. I.

1966. Biology and systematics of American species of the genus *Hesperapis* Cockerell. Ph.D. Diss., Univ. Calif., Berkeley, 461 pp.

THORP, R. W.
1969. Systematics and ecology of bees of the subgenus *Diandrena* (Hymenoptera: Andrenidae). Univ. Calif. Publ. Entomol. 52:1–146.

TIMBERLAKE, P. H.
1937. New species of *Andrena* from California. Pan-Pac. Entomol. 13(1–2):69–74.

TURNER, W. J.
1966. Behavior and activities during foraging periods of *Lasioglossum* (*Sphecodogastra*) *aberrans* (Crawford) at Antioch, California (Hymenoptera, Halictidae). M.S. thesis, Univ. Calif., Berkeley, 126 pp.

PLATES

Upper left: *Andrena (Diandrena) chalybaea* female gathering pollen from flowers of *Camissonia ovata*. Upper right: *Andrena (Diandrena) foxii* female gathering pollen from flowers of *Camissonia campestris.* Lower: *Andrena (Diandrena) foxii* male resting on soil inside "territory" among flowers of *Camissonia campestris.*

Upper: *Andrena (Diandrena) macswaini* aggregated nest tumuli in small runoff gulley. Lower left: *Andrena (Diandrena) foxii* plaster cast of L-shaped burrow terminating in a cell. Lower right: *Andrena (Diandrena) macswaini* plaster cast of deep burrow, cells not located.

Upper: *Camissonia cheiranthifolia cheiranthifolia* habitat near Fleishhacker Zoo, San Francisco.
Lower: *Oenothera hookeri* habitat near Fleishhacker Zoo, San Francisco.

Upper: *Camissonia campestris* habitat near Tipton. Lower: *Camissonia sierrae sierrae* habitat near Oakhurst.

Upper: *Camissonia ovata* habitat in El Cerrito, Berkeley Hills, spring 1962. Lower: El Cerrito *C. ovata* habitat in spring 1964 excavated for house construction.

Upper: *Camissonia campestris*. Middle: *Camissonia cheiranthifolia cheiranthifolia* at Point Reyes. Lower left: *Camissonia boothii decorticans* near Bakersfield. Lower right: *Oenothera deltoides cognata* near Livingston.